GUIDED NOTEBOOK

MYMATHLAB®
PRECALCULUS:
A UNIT CIRCLE APPROACH

Kirk Trigsted

University of Idaho

PEARSON

Boston Columbus Indianapolis New York San Francisco Upper Saddle River
Amsterdam Cape Town Dubai London Madrid Milan Munich Paris Montreal Toronto
Delhi Mexico City Sao Paulo Sydney Hong Kong Seoul Singapore Taipei Tokyo

Copyright © 2012 Pearson Education, Inc.
Publishing as Pearson, 75 Arlington Street, Boston, MA 02116.

ISBN-13: 978-0-321-73677-2
ISBN-10: 0-321-73677-X

4 5 6 V031 16 15 14

www.pearsonhighered.com

Table of Contents

Section 1.1 Guided Notebook

Section 1.1 The Rectangular Coordinate System
Work through Section 1.1 TTK #1
Work through Objective 1
Work through Objective 2
Work through Objective 3
Work through Objective 4

Section 1.1 The Rectangular Coordinate System

1.1 Things To Know

1. Simplifying Radicals (Appendix A.3)

Make sure that you can simplify radicals such as $\sqrt{108}$. Can you show that the expression $\sqrt{108}$ is equivalent to $6\sqrt{3}$?

Read the introduction to Section 1.1 and write notes here:

1

Section 1.1

Draw a rectangular coordinate system and label the four quadrants here:

Section 1.1 Objective 1 Plotting Ordered Pairs

Work through the video that accompanies Example 1 and write your notes here:

Plot the ordered pairs (-2,3),(0,4),(2,5) and (4,6) and state in which quadrant or on which axis each pair lies.

Section 1.1 Objective 2 Graphing Equations by Plotting Points

Work through Example 2 writing your notes here:

Sketch the graph of $y = x^2$.

Work through the video that accompanies Example 3 and write your notes here:

2

Determine whether the following ordered pairs lie on the graph of the equation $x^2 + y^2 = 1$.

a) $(0, -1)$ b) $(1, 0)$ c) $\left(\frac{1}{3}, \frac{2}{3} \right)$ d) $\left(-\frac{\sqrt{2}}{2}, \frac{\sqrt{2}}{2} \right)$

Section 1.1

Section 1.1 Objective 3 Finding the Midpoint of a Line Segment Using the Midpoint Formula

Write down the **midpoint formula** here:

Work through Example 4:

Find the midpoint of the line segment whose endpoints are $(-3, 2)$ and $(4, 6)$.

Work through the video that accompanies Example 5 and write your notes here:

In geometry, it can be shown that four points in a plane form a parallelogram if the two diagonals of the quadrilateral formed by the four points bisect each other. Do the points $A(0, 4)$, $B(3, 0)$, $C(9, 1)$, and $D(6, 5)$ form a parallelogram?

4

Section 1.1 Objective 4 Finding the Distance Between Two Points Using the Distance Formula

Watch the video that accompanies Objective 4. Take notes below.

Write the **distance formula** here:

Work through Example 6:

Find the distance between the points $A(-1,5)$ and $B(4,-5)$.

Work through the video that accompanies Example 7 and write your notes here:

Verify that the points $A(3,-5)$, $B(0,6)$, and $C(5,5)$ form a right triangle.

Section 1.2 Guided Notebook

Section 1.2 Circles

Work through Section 1.2 TTK #1
Work through Section 1.2 TTK #3
Work through Section 1.2 TTK #4
Work through Objective 1
Work through Objective 2
Work through Objective 3

Section 1.2 Circles

1.2 Things To Know

1. Solving Quadratic Equations by Completing the Square (Appendix B.4)
    Review this objective as necessary. You MUST understand the concept of completing the square before you can be successful working with circles. What number must you add to complete the square: $x^2 - 10x$?

    You should get an answer of 25. Why?

3. Finding the Midpoint of a Line Segment (Section 1.1)
    Do you remember the Midpoint Formula? Write the midpoint formula here:

    Midpoint Formula:

What is the midpoint of the line segment joining points A and B? $A(4,-1)$; $B(5,6)$

You should get a midpoint of $\left(\dfrac{9}{2}, \dfrac{5}{2}\right)$.

4. Finding the Distance between Two Points Using the Distance Formula (Section 1.1)
   Do you remember the Distance Formula? Write the distance formula here:

   Distance Formula:

What is the distance $d(A, B)$ between the points A and B? $A(10,3)$; $B(-2,19)$.

You should get a distance of 20.

An Introduction to Circles
Read the introduction to Section 1.2.

What is the definition of a circle?

See how to derive the equation of a circle by watching the **animation** found in the introduction and take notes here:

Write the **standard form of an equation of a circle** here:

Section 1.2

Section 1.2 Objective 1 Writing the Standard Form of an Equation of a Circle
Work through the video that accompanies Example 1 and take notes here:

Find the standard form of the equation of the circle whose center is $(-2,3)$ and with

radius 6.

Work through Example 2 and take notes here:

Find the standard form of a circle whose center is $(0,6)$ and that passes through the

point $(4,2)$.

Work through the video that accompanies Example 3 and take notes here:

Find the standard form of the equation of the circle that contains endpoints of a diameter at $(-4,-3)$ and $(2,-1)$.

Section 1.2

Section 1.2 Objective 2 Sketching the Graph of a Circle

Work through the video that accompanies Example 4 and take notes here:

Find the center and radius, and sketch the graph of the circle $(x-1)^2 + (y+2)^2 = 9$.

Also find any intercepts. **(Be sure to pay close attention to how we find intercepts.)**

Note: When determining intercepts algebraically, what does it mean if you get an imaginary solution?

<u>Section 1.2 Objective 3 Converting the General Form of a Circle into Standard Form</u>

What is the difference between standard form and general form?

Work through the video that accompanies Example 5 and take notes here:

Write the equation $x^2 + y^2 - 8x + 6y + 16 = 0$ in standard form; find the center, radius, and intercepts, and sketch the graph.

Carefully work through the animation that accompanies Example 6 and take notes here:

Write the equation $4x^2 + 4y^2 + 4x - 8y + 1 = 0$ in standard form; find the center, radius, and intercepts, and sketch the graph.

Section 1.3 Guided Notebook

Section 1.3 Lines
Work through Objective 1
Work through Objective 2
Work through Objective 3
Work through Objective 4
Work through Objective 5
Work through Objective 6
Work through Objective 7
Work through Objective 8

Section 1.3 Lines

Section 1.3 Objective 1 Determining the Slope of a Line

Watch the video that accompanies Objective 1.
Write down the **definition of slope** here:

Section 1.3

Work through Example 1: Find the slope of the line that passes through the indicated ordered pairs.

    a) $(-2,3)$ and $(2,-5)$ b) $(6,-4)$ and $(-5,1)$

Section 1.3 Objective 2 Sketching a Line Given a Point and the Slope

Work through the video that accompanies Example 2 and take notes here:

    Sketch the line with slope $m = \frac{2}{3}$ that passes through the point $(-1,-4)$. Also find three more points located on the line.

16

Section 1.3 Objective 3 Finding the Equation of a Line Using the Point-Slope Form
Watch the video that accompanies Objective 3 to see how to derive the Point-Slope Form of the equation of a line and take notes here:

Write down the **Point-Slope Form** of a line here:

Work through Example 3 and take notes here:
Find an equation in point-slope form of the line with slope $m = \frac{2}{3}$ that passes through the point $(-1, -4)$.

Section 1.3

<u>Section 1.3 Objective 4 Finding the Equation of a Line Using the Slope-Intercept Form</u>
In Example 3 from the previous page, you should have found the equation of the line to be
$y + 4 = \dfrac{2}{3}(x + 1)$. Try solving this equation for y. What do you get?

Now, write down the **Slope-Intercept Form** of the equation of a line here:

Work through Example 4 and take notes here:

Find the equation of the line with slope $\dfrac{1}{4}$ and y-intercept 3, and write your answer in

slope-intercept form.

18

Section 1.3 Objective 5 Writing the Equation of a Line in Standard Form
Write down the **Standard Form Equation of a Line** here:

Work through the video that accompanies Example 5 and take notes here:

    Find the equation of the line passing through the points $(-1, 3)$ and $(2, -4)$. Write the

    equation in point-slope form, slope-intercept form, and standard form.

Section 1.3

Make sure that you know how to write an equation of a line in point-slope form, slope-intercept form, and standard form! Write the point-slope, slope-intercept, and standard forms here:

Point-Slope Form:

Slope-Intercept Form:

Standard Form:

Section 1.3 Objective 6 Finding the Slope and y-Intercept of a Line in Standard Form
Watch the video that accompanies Objective 6 and take notes here:

Given a line of the form $Ax + By = C$, $B \neq 0$ what is the slope of this line and what is the y-intercept?

Slope = _____ and y-intercept = _____

Now work through Example 6 and write your notes here:

Find the slope and y-intercept and sketch the line $3x - 2y = 6$.

<u>Section 1.3 Objective 7 Sketching Lines by Plotting Intercepts</u>

Watch the video that accompanies Example 7 and take notes here:

Sketch the line $2x - 5y = 8$ by plotting intercepts.

What is the definition of an ***x*-intercept**?

Section 1.3

What is the definition of a **y-intercept**?

Section 1.3 Objective 8 Finding the Equations of Horizontal and Vertical Lines
Horizontal Lines: Watch the video that describes the equation of a horizontal line and take notes here:

What is the slope of every horizontal line?

What is the equation of a horizontal line?

Vertical Lines: Watch the video that describes the equation of a vertical line and take notes here:

Does a vertical line have slope?

What is the equation of a vertical line?

Work through Example 8 and take notes here:

    a) Find the equation of the horizontal line passing through the point $(-1,3)$.

    b) Find the equation of the vertical line passing through the point $(-1,3)$.

Section 1.3

Before going on to Section 2.4, you may want to write all of the different types of equations of lines for future reference. These forms are summarized at the end of Section 1.3 in your eText.

Point-Slope Form

Slope-Intercept Form

Standard Form

Horizontal Line

Vertical Line

24

Section 1.4 Guided Notebook

Section 1.4 Parallel and Perpendicular Lines
    Work through Objective 1
    Work through Objective 2
    Work through Objective 3
    Work through Objective 4
    Work through Objective 5

Section 1.4 Parallel and Perpendicular Lines

<u>Section 1.4 Objective 1 Understanding the Definition of Parallel Lines</u>
Write down the Theorem found in Objective 1:

Work through Example 1 and take notes here:
    Show that the lines $y = -\dfrac{2}{3}x - 1$ and $4x + 6y = 12$ are parallel.

Section 1.4

<u>Section 1.4 Objective 2 Understanding the Definition of Perpendicular Lines</u>
Write down the Theorem found in Objective 2:

Draw and label Figure 29 here:

Work through Example 2 and take notes here:

Show that the lines $3x - 6y = -12$ and $2x + y = 4$ are perpendicular.

Write down the Summary of Parallel and Perpendicular Lines following Example 2 here:

Section 1.4 Objective 3 Determining Whether Two Lines Are Parallel, Perpendicular, or Neither

Watch the video that accompanies Example 3 and take notes here:

> For each of the following pairs of lines, determine whether the lines are parallel, perpendicular, or neither.

a) $3x - y = 4$
$x + 3y = 7$

b) $y = \dfrac{1}{2}x + 3$
$x + 2y = 1$

c) $x = -1$
$x = 3$

Section 1.4

<u>Section 1.4 Objective 4 Finding the Equations of Parallel and Perpendicular Lines</u>
You may want to turn back to your notes from Section 1.3 and write down the following
equations of lines:

Point-Slope Form

Slope-Intercept Form

Standard Form

Horizontal Line

Vertical Line

Watch the video that accompanies Example 4 and take notes here:
Find the equation of the line parallel to the line $2x + 4y = 1$ that passes through the
point $(3, -5)$. Write the answer in point-slope form, slope-intercept form, and
standard form.

Watch the video that accompanies Example 5 and take notes here:

Find the equation of the line perpendicular to the line $y = -5x + 2$ that passes through the point $(3, -1)$. Write the answer in slope-intercept form.

Section 1.4

Section 1.4 Objective 5 Solving a Geometric Application of Parallel and Perpendicular
Lines.

What two methods can be used to determine if 4 points in a plane form a parallelogram?

If you have a parallelogram, how can you use slopes to determine if it is a rhombus?

Watch the video that accompanies Example 5 and take notes here:

Do the points A(0,4), B(3,0), C(9,1), and D(6,5) form a parallelogram? If the points
form a parallelogram, then is the parallelogram a rhombus?

Section 2.1 Guided Notebook

Section 2.1 Relations and Functions

Work through Section 2.1 TTK #3
Work through Section 2.1 TTK #4
Work through Objective 1
Work through Objective 2
Work through Objective 3
Work through Objective 4
Work through Objective 5

Section 2.1 Relations and Functions

2.1 Things To Know

3. Solving Polynomial Inequalities (Appendix B.9)

Review this objective as necessary. You will have to be able to solve a polynomial inequality later on during this homework assignment.

Can you solve the inequality $2x^2 + 5x - 3 \geq 0$? You should get a solution of $(-\infty, -3] \cup [\frac{1}{2}, \infty)$. Watch the appropriate video and review Appendix B.9 as needed.

Section 2.1

4. Solving Rational Inequalities (Appendix B.9)

Review this objective as necessary. You will have to be able to solve a rational inequality later on during this homework assignment.

Can you solve the inequality $\dfrac{x^2 - 3x - 10}{x + 4} \geq 0$? You should get a solution of

$(-4, -2] \cup [5, \infty)$. Watch the appropriate video and review Appendix B.9 as needed.

Section 2.1 Objective 1 Understanding the Definitions of Relations and Functions

Watch the video that accompanies the definition of a relation and take notes here:

Write down the definition of a **relation.**

Write an example of a relation as seen in the video and state the domain and range of the relation.

32

Write down the definition of a **function.**

Read through the text preceding Example 1 to further clarify functions and relations and take notes here.

Work through the video that accompanies Example 1 and take notes here:

Determine whether each relation is a function, and then find the domain and range.

a.

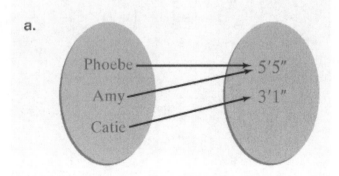

b.

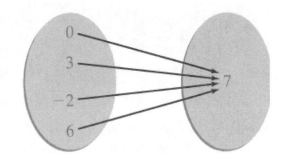

c. $\{(3, 7), (-3, 2), (-4, 5), (1, 4), (3, -4)\}$

d. $\{(-1, 2), (0, 2), (1, 5), (2, 3), (3, 2)\}$

Work through Example 2 and take notes here: Use the domain and range of the following relation to determine whether the relation is a function.

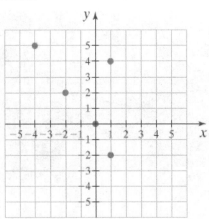

Section 2.1 Objective 2 Determining Whether Equations Represent Functions
Work through the interactive video that accompanies Example 3 and take notes here:

Section 2.1

Section 2.1 Objective 3 Using Function Notation; Evaluating Functions

Carefully take notes from the video that accompanies Objective 3.

Fill out the following table as seen in this video. (Add more values if you wish.)

Domain Value	Range Value	Ordered Pair
x	$f(x) = \dfrac{3}{2}x + 6$	$(x, f(x))$

Now sketch the function $f(x) = \dfrac{3}{2}x + 6$ using the values from the table on the previous page. (Also, see Figure 3 in your eText.)

What is the definition of an **independent variable**?

What is the definition of a **dependent variable**?

Work through the interactive video that accompanies Example 4 and take notes here:

Rewrite these equations using function notation where y is a function of x. Then answer the question following each equation.

a. $3x - y = 5$

What is the value of $f(4)$?

b. $x^2 - 2y + 1 = 0$

Does the point (-2, 1) lie on the graph of this function?

c. $y + 7 = 0$

What is $f(x)$ when $x = 3$?

Work through Example 5:

Given that $f(x) = x^2 + x - 1$, evaluate the following:

a) $f(0)$ b) $f(-1)$ c) $f(x+h)$ d) $\dfrac{f(x+h) - f(x)}{h}$

Section 2.1 Objective 4 Using the Vertical Line Test
Watch the video that accompanies Objective 4 and write your notes here:

In your own words, explain how the **vertical line test** works:

Work through Example 6 in your eText and take notes here: Use the vertical line test to determine which of the following graphs represents the graph of a function.

a.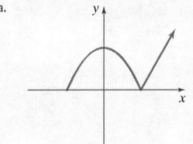

b.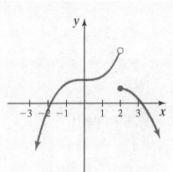

c.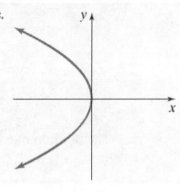

Section 2.1 Objective 5 Determining the Domain of a Function Given the Equation
Carefully work through the video that accompanies Objective 5 and fill in the blanks seen on the following three pages:

The **domain** of a function $y = f(x)$ is ...

Define a **polynomial function** here:

Give an example of a polynomial function here:

What is the domain of every polynomial function?

41

Define a **rational function** here:

The **domain** of a rational function is…

Give an example of a rational function here and find the domain.

Define a **root function** here:

Write the two examples of root functions seen in this video and find the domains of each.

Below is a summary of the three classifications of functions seen in the previous video and a summary of how to find the domain of each. Use this summary for future reference.

Class of Function	Form	Domain
Polynomial functions	$f(x) = a_n x^n + a_{n-1} x^{n-1} + \cdots + a_1 x + a_0$	Domain is $(-\infty, \infty)$.
Rational functions	$f(x) = \dfrac{g(x)}{h(x)}$ where g and h are polynomial functions such that $h(x) \neq 0$	Domain is all real numbers such that $h(x) \neq 0$.
Root functions	$f(x) = \sqrt[n]{g(x)}$, where $g(x)$ is a function and n is an integer such that $n \geq 2$.	1. If n is even, the domain is the solution to the inequality $g(x) \geq 0$. 2. If n is odd, the domain is the set of all real numbers for which g is defined.

Section 2.1

Work through interactive video that accompanies Example 7 and explain how to find the domain of each of the following functions:

a) $f(x) = 2x^2 - 5x$

b) $f(x) = \dfrac{x}{x^2 - x - 6}$

c) $h(x) = \sqrt{x^2 - 2x - 8}$

d) $f(x) = \sqrt[3]{5x - 9}$

b) $k(x) = \sqrt[4]{\dfrac{x-2}{x^2+x}}$

Section 2.2 Guided Notebook

Section 2.2 Properties of a Function's Graph
        Work through Objective 1
        Work through Objective 2
        Work through Objective 3
        Work through Objective 4
        Work through Objective 5
        Work through Objective 6

Section 2.2 Properties of a Function's Graph

<u>2.2 Things To Know</u>

Take a few moments to work through each of the four Things to Know objectives. You must have a solid understanding of each of these topics in order to fully understand the material presented in Section 2.2. You may want to work through each of the "You Try It" problems and review the material as needed before going on. Take notes here if necessary:

Section 2.2 Objective 1 Determining the Intercepts of a Function

Work through the video that accompanies Objective 1 and fill in the notes below:

Define **y-intercept**:

Define **x-intercept**:

Given a function *f(x)*, how can we find the intercepts?

Show how to find the intercepts of the functions $f(x) = -3x + 2$ and $g(x) = x^3 - x^2 - 12x$.

Section 2.2

Work through Example 2:

Find all intercepts of the function $f(x) = x^3 - 2x^2 + x - 2$.

Section 2.2 Objective 2 Determining the Domain and Range of a Function from Its Graph

Watch the video that accompanies Objective 2 and take notes here:

The examples seen in the video are the **same** graphs used in Example 3:
Find the domain and range of the functions seen below:

a.

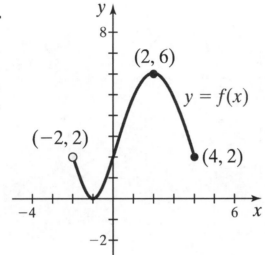

Domain:

Range:

b.

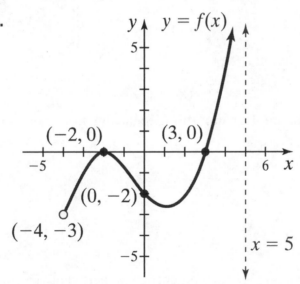

Domain:

Range:

c.

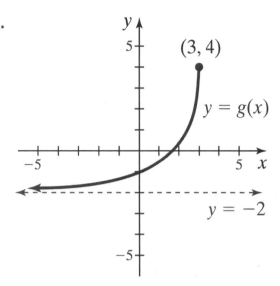

Domain:

Range:

What is the definition of a **vertical asymptote**? (See the link in the solution to Example 3 part b.)

What is the definition of a **horizontal asymptote**? (See the link in the solution to Example 3 part c.)

Section 2.2 Objective 3 Determining Whether a Function is Increasing, Decreasing, or Constant

Watch the video that accompanies Objective 3 and fill in the following definitions:
   Increasing:

Decreasing:

Constant

Watch the video that accompanies Example 4 and determine the intervals on which the function seen below is increasing, decreasing, and constant: (**THE INTERVAL(S) ARE ALWAYS *X*-INTERVALS.**)

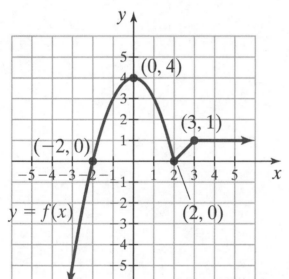

The function is increasing on the interval(s):

The function is decreasing on the interval(s):

The function is constant on the interval(s):

Section 2.2

Section 2.2 Objective 4 Determining Relative Maximum and Relative Minimum Values of a Function

Watch the video that accompanies Objective 4 and fill in the following definitions:

Relative Maximum:

Relative Minimum:

Work through Example 5 and answer each of the following questions:

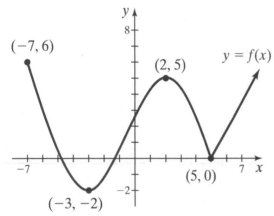

a)   On what interval(s) is f increasing?

b) On what interval(s) is f decreasing?

c) For what value(s) of x does f have a relative minimum?

d) For what value(s) of x does f have a relative maximum?

e) What are the relative minima?

f) What are the relative maxima?

Section 2.2 Objective 5 Determining Whether a Function is Even, Odd, or Neither

Watch the video that accompanies Objective 5 and fill in the following definitions:

Even Functions:

(Draw an example of an even function here.)

Odd Functions:

(Draw an example of an odd function here.)

Section 2.2

Summarize Even and Odd Functions by filling in the following Table.

TYPE OF FUNCTION	ALGEBRAIC DEFINITION	TYPE OF SYMMETRY (Write y-axis or origin symmetry.)
Even	$f(-x) = $ _____	
Odd	$f(-x) = $ _____	

Work through Example 6:

Determine whether each function is even, odd, or neither. (Explain in your own words why each function is even, odd, or neither.)

a.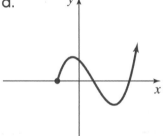

b.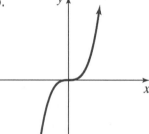

c.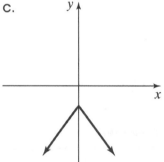

You do NOT need the graph of a function to determine if it is even, odd, or neither! Carefully, work through the video that accompanies Example 7 and determine WITHOUT GRAPHING whether each of the following functions is even, odd, or neither:

a) $f(x) = x^3 + x$

b) $g(x) = \dfrac{1}{x^2} + 7|x|$

c) $h(x) = 2x^5 - \dfrac{1}{x}$

d) $G(x) = x^2 + 4x$

Section 2.2

Section 2.2 Objective 6 Determining Information about a Function from a Graph
Work through the animation that accompanies Example 8 and answer each question:

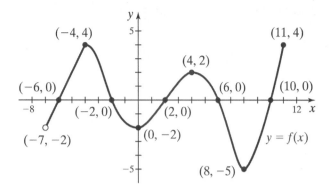

a) What is the y-intercept?

b) What are the real zeros of *f*?

c) Determine the domain and range of *f*.

d) Determine the interval(s) on which *f* is increasing, decreasing and constant.

e) For what value(s) of *x* does *f* obtain a relative maximum? What are the relative maxima?

f) For what value(s) of *x* does *f* obtain a relative minimum? What are the relative minima?

g) Is *f* even, odd or neither?

h) For what values of *x* is $f(x) = 4$?

i) For what values of *x* is $f(x) < 0$?

56

Section 2.3 Guided Notebook

Section 2.3 Graphs of Basic Functions; Piecewise Functions
Work through Objective 1
Work through Objective 2
Work through Objective 3

Section 2.3 Graphs of Basic Functions; Piecewise Functions

Section 2.3 Objective 1 Sketching the Graphs of the Basic Functions
Read through all the text in Objective 1. **YOU MUST MEMORIZE THE GRAPHS** of the
9 basic functions seen on these pages! On the next three pages of this notebook, sketch each
of these functions and list the properties of each. Click on the appropriate link in your eText
to see the properties of each function.

1. The constant function $f(x) = b$

2. The identity function $f(x) = x$

3. The square function $f(x) = x^2$

4. The cube function $f(x) = x^3$

5. The absolute value function $f(x) = |x|$

6. The square root function $f(x) = \sqrt{x}$

7. The cube root function $f(x) = \sqrt[3]{x}$

8. The reciprocal function $f(x) = \dfrac{1}{x}$

Section 2.3

9. The greatest integer function $f(x) = [\![x]\!]$

Section 2.3 Objective 2 Analyzing Piecewise-Defined Functions
Read through the text preceding Example 1 taking notes here:

Carefully work through the **animation** that accompanies Example 1 and take notes here: (You should use a pencil to sketch piecewise functions. Why? Watch the animation to find out.)

Sketch the function $f(x) = \begin{cases} x^2 & \text{if } x < 1 \\ 1-x & \text{if } x \geq 1 \end{cases}$

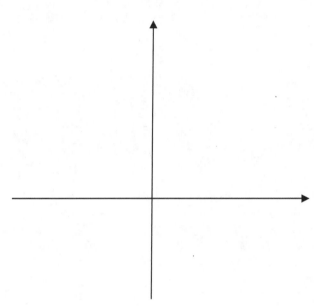

Section 2.3

Work through the video that accompanies Example 2 and take notes here:

$$\text{Let } f(x) = \begin{cases} 1 & \text{if } x < -1 \\ \sqrt[3]{x} & \text{if } -1 \le x < 0 \\ \dfrac{1}{x} & \text{if } x > 0 \end{cases}$$

a) Evaluate $f(-3)$, $f(-1)$, and $f(2)$.

b) Sketch the graph of f.

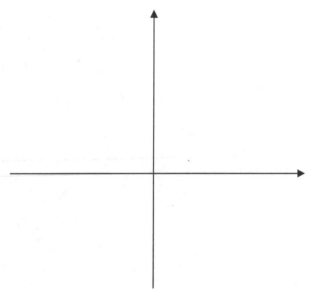

c) Determine the domain of f.

d) Determine the range of f.

Work through the video that accompanies Example 3 and take notes here:

Find a rule that describes the following piecewise function:

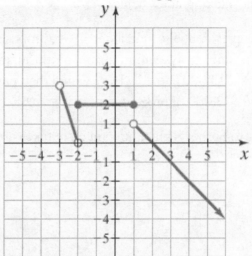

Section 2.3

Section 2.3 Objective 3 Solving Applications of Piecewise-Defined Functions

Work through Example 4 taking notes here:

> Steve Forbes, a presidential candidate in 1996, proposed a flat tax to replace the existing U.S. income tax system. In his tax proposal, every adult would pay $0.00 in taxes on the first $13,000 earned. They would then pay a flat tax of 17% on everything over $13,000. Forbes' tax plan is actually a piecewise-defined function.
>
> a. According to Forbes's plan, how much in taxes are owed for someone earning $50,000?

> b. Find the piecewise function, $T(x)$, that describes the amount of taxes paid, T, as a function of the dollars earned, x, for Forbes' tax plan.

c. Sketch the piecewise function, $T(x)$.

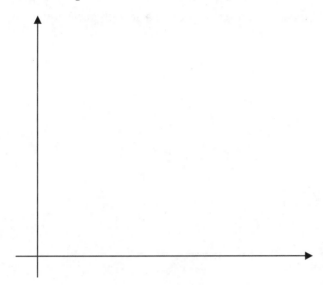

Work through the video that accompanies Example 5 taking notes here:

Cheapo Rental Car Co. charges a flat rate of $85 to rent a car for up to 2 days. The company charges an additional $20 for each additional day (or part of a day).

a. How much does it cost to rent a car for 2 ½ days? 3 days?

b. Write and sketch the piecewise function, $C(x)$, that describes the cost of renting a car as a function of time, x, in days rented for values of x less than or equal to 5.

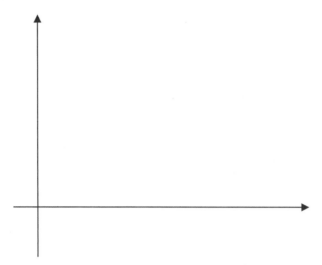

Section 2.4 Transformations of Functions

Work through Section 2.4 TTK #1
Work through Section 2.4 TTK #4
Work through Objective 1
Work through Objective 2
Work through Objective 3
Work through Objective 4
Work through Objective 5
Work through Objective 6

Section 2.4 Transformations of Functions

2.4 Things To Know

1. Determining the Domain of a Function Given the Equation (Section 2.1)

You may want to work through the interactive video to remember how to find the domain of each of these functions.

a) $f(x) = 2x^2 - 5x$ b) $f(x) = \dfrac{x}{x^2 - x - 6}$ c) $h(x) = \sqrt{x^2 - 2x - 8}$

d) $f(x) = \sqrt[3]{5x - 9}$ e) $k(x) = \sqrt[4]{\dfrac{x-2}{x^2 + x}}$

Section 2.4

4. Sketching the Graphs of the Basic Functions

   You MUST have the graphs of the following functions memorized before starting Section 2.4. Can you easily sketch the graphs of these 8 functions? You can review these in Section 2.3.

    1. The constant function $f(x) = b$

    2. The identity function $f(x) = x$

    3. The square function $f(x) = x^2$

    4. The cube function $f(x) = x^3$

    5. The absolute value function $f(x) = |x|$

    6. The square root function $f(x) = \sqrt{x}$

    7. The cube root function $f(x) = \sqrt[3]{x}$

    8. The reciprocal function $f(x) = \dfrac{1}{x}$

Section 2.4 Objective 1 Using Vertical Shifts to Graph Functions
Work through the video that accompanies Example 1 and take notes here:
Sketch the graphs of $f(x) = |x|$ and $g(x) = |x| + 2$.

If $c > 0$, explain in your own words how to sketch the graph of $y = f(x) + c$ and
$y = f(x) - c$. Click on the "**animate buttons**" on p. 2.4-6 and use the given graph of
$y = f(x)$ below to sketch the graphs of $y = f(x) + c$ and $y = f(x) - c$.

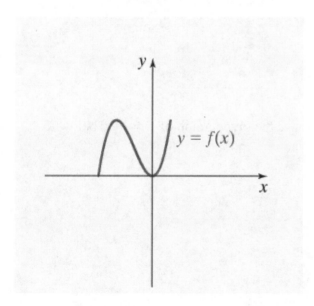

Section 2.4

Section 2.4 Objective 2 Using Horizontal Shifts to Graph Functions
Work through the video that accompanies Objective 2 and take notes here:

If $c > 0$, explain in your own words how to sketch the graph of $y = f(x+c)$ and
$y = f(x-c)$.

Click on the "**animate buttons**" on p. 2.4-8 and use the given graph of $y = f(x)$ below to
sketch the graphs of $y = f(x+c)$ and $y = f(x-c)$.

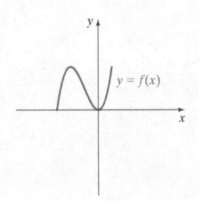

IT IS NOW TIME TO COMBINE A HORIZONTAL AND A VERTICAL SHIFT.

Work through the animation that accompanies Example 2 and take notes here:

Use the graph of $y = x^3$ to sketch the graph of $g(x) = (x-1)^3 + 2$.

Section 2.4 Objective 3 Using Reflections to Graph Functions

Work through the video that accompanies Objective 3 and take notes here:

71

If $c > 0$, explain in your own words how to sketch the graph of $y = -f(x)$ and $y = f(-x)$.

Click on the "**animate buttons**" on p. 2.4-11 and p. 2.4-12 and use the given graphs of $y = f(x)$ below to sketch the graphs of $y = -f(x)$ and $y = f(-x)$.

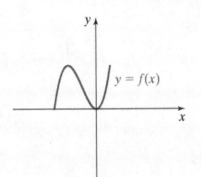

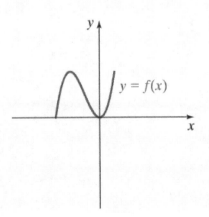

Work through Example 3 and take notes here: Use the graph of the basic function $y = \sqrt[3]{x}$ to sketch each graph.

a) $y = -\sqrt[3]{x} - 2$

b) $y = \sqrt[3]{1-x}$

Section 2.4

Section 2.4 Objective 4 Using Vertical Stretches and Compressions to Graph Functions
Work through the video that accompanies Example 4 and take notes here:

If $a > 1$, explain in your own words how to sketch the graph of $y = af(x)$.

If $0 < a < 1$, explain in your own words how to sketch the graph of $y = af(x)$.

74

Click on the **"animate buttons"** on p. 2.4-17 and use the given graph of $y = f(x)$ below to sketch the graphs of $y = af(x)$ for $a > 1$ and $y = af(x)$ for $0 < a < 1$.

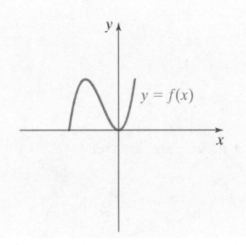
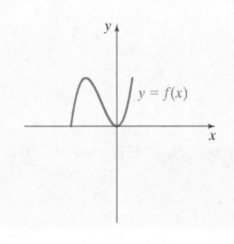

Section 2.4 Objective 5 Using Horizontal Stretches and Compressions to Graph Functions
Work through the video that accompanies Objective 5 and take notes here:

If $a > 1$, explain in your own words how to sketch the graph of $y = f(ax)$.

If $0 < a < 1$, explain in your own words how to sketch the graph of $y = f(ax)$.

Click on the "**animate buttons**" on p. 2.4-19 and use the given graph of $y = f(x)$ below to sketch the graphs of $y = f(ax)$ for $a > 1$ and $y = f(ax)$ for $0 < a < 1$.

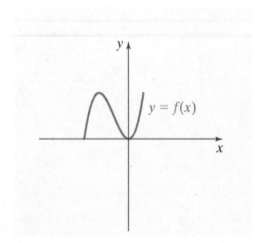
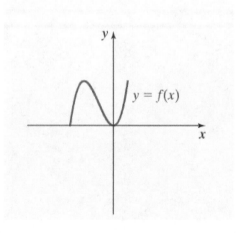

Section 2.4 Objective 6 Using Combinations of Transformations to Graph Functions
You have learned 6 transformations in this section. You may encounter functions that combine many (if not all) of these transformations. Write down the "order of operations" of transformations as in Objective 6.

"Order of Operations" for Transformations

1.

2.

3.

4.

5.

6.

Work through the animation that accompanies Example 6 and take notes here:

Use transformations to sketch the graph of $y = -2(x+3)^2 - 1$.

Section 2.4

Work through the interactive video that accompanies Example 7 and take notes here:
Use the graph of $y = f(x)$ to sketch each of the following functions.

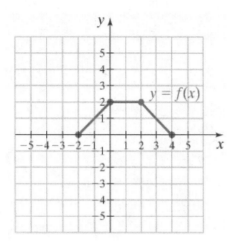

a) $y = -f(2x)$ b) $y = 2f(x-3) - 1$ c) $y = -\dfrac{1}{2}f(2-x) + 3$

Use the **Summary of Transformation Techniques** to complete the following statements:

Given a function $y = f(x)$ and a constant $c > 0$:

1. The graph of $y = f(x) + c$ is obtained by...

2. The graph of $y = f(x) - c$ is obtained by...

3. The graph of $y = f(x + c)$ is obtained by...

4. The graph of $y = f(x - c)$ is obtained by...

5. The graph of $y = -f(x)$ is obtained by...

6. The graph of $y = f(-x)$ is obtained by...

7. Suppose a is a positive real number. The graph of $y = af(x)$ is obtained by...

8. Suppose a is a positive real number. The graph of $y = f(ax)$ is obtained by...

Section 2.5 The Algebra of Functions; Composite Functions
   Work through Objective 1
   Work through Objective 2
   Work through Objective 3
   Work through Objective 4
   Work through Objective 5

Section 2.5 The Algebra of Functions; Composite Functions

Work through the video in the introduction and learn about the Algebra of Functions.

   Write down 4 definitions below and give one example of each:

   1. The sum of f and g

   2. The difference of f and g

   3. The product of f and g

   4. The quotient of f and g

Section 2.5 Objective 1 Evaluating a Combined Function
Work through Example 1 and take notes here:

$$\text{Let } f(x) = \frac{12}{2x+4} \text{ and } g(x) = \sqrt{x}. \text{ Find each of the following:}$$

a) $(f+g)(1)$ b) $(f-g)(1)$ c) $(fg)(4)$ d) $\left(\dfrac{f}{g}\right)(4)$

Work through the interactive video that accompanies Example 2 and take notes here:
Use the graph to evaluate each expression or state that it is undefined:

a) $(f+g)(1)$

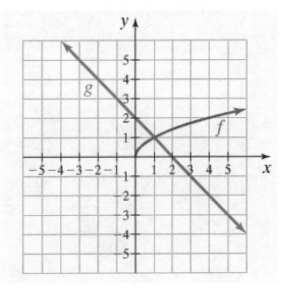

b) $(f-g)(0)$

c) $(fg)(4)$

d) $\left(\dfrac{f}{g}\right)(2)$

81

Section 2.5

Section 2.5 Objective 2 Finding the Intersection of Intervals
In order to find the domain of a combined function, you MUST be able to find the intersection of two or more intervals. Watch the video that accompanies Example 3 and take notes here:

Find the intersection of the following intervals:

a) $[0,\infty)\cap(-\infty,5]$

b) $((-\infty,-2)\cup(-2,\infty))\cap[-4,\infty)$

Section 2.5 Objective 3 Finding Combined Functions and Their Domains

Work through the video that accompanies Objective 3 and take notes here:

Section 2.5

Work through the video that accompanies Example 4 and take notes here:

Let $f(x) = \dfrac{x+2}{x-3}$ and $g(x) = \sqrt{4-x}$. Find:

    a) $f + g$, b) $f - g$, c) fg, and d) $\dfrac{f}{g}$ and the domain of each.

Section 2.5 Objective 4 Forming and Evaluating Composite Functions
Carefully work through the video that accompanies Objective 4 and take notes here. **MAKE SURE THAT YOU CAN DEFINE THE COMPOSITE FUNCTION:**

Work through the interactive video that accompanies Example 5 and take notes here:

Let $f(x) = 4x+1$, $g(x) = \dfrac{x}{x-2}$ and $h(x) = \sqrt{x+3}$.

a) Find the function $f \circ g$.

b) Find the function $g \circ h$.

c) Find the function $h \circ f \circ g$.

d) Evaluate $(f \circ g)(4)$, or state that it is undefined.

e) Evaluate $(g \circ h)(1)$, or state that it is undefined.

f) Evaluate $(h \circ f \circ g)(6)$, or state that it is undefined.

Work through the interactive video that accompanies Example 6 and take notes here:
Use the graph to evaluate each expression or state that it is undefined:

a) $(f \circ g)(4)$

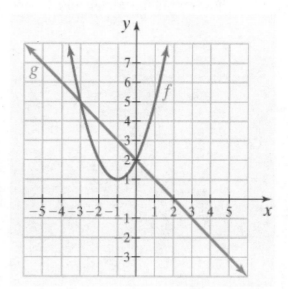

b) $(g \circ f)(-3)$

c) $(f \circ f)(-1)$

d) $(g \circ g)(4)$

e) $(f \circ g \circ f)(1)$

Section 2.5

Section 2.5 Objective 5 Determining the Domain of Composite Functions

Work through the video that accompanies Objective 5 and take notes here.

Carefully write down how to find the domain of a composite function and give an example of how to find the domain of a composite function.

Work through the interactive video that accompanies Example 7 and take notes here.

Let $f(x) = \dfrac{-10}{x-4}$, $g(x) = \sqrt{5-x}$, and $h(x) = \dfrac{x-3}{x+7}$.

a) Find the domain of $f \circ g$.

b) Find the domain of $g \circ f$.

c) Find the domain of $f \circ h$.

d) Find the domain of $h \circ f$.

Section 2.6 Guided Notebook

Section 2.6 One-to-one Functions; Inverse Functions
    Work through Objective 1
    Work through Objective 2
    Work through Objective 3
    Work through Objective 4
    Work through Objective 5

Section 2.6 One-To-One Functions; Inverse Functions

Section 2.6 Objective 1 Understanding the Definition of a One-to-One Function
Work through the video that accompanies Objective 1 and write notes here.

Write down the definition of a **one-to-one function:**

Give an example of a function that is one-to-one.

Give an example of a function that is **not** one-to-one.

Write down the **alternate definition of a one-to-one function** as seen in the eText.

Section 2.6 Objective 2 Determining Whether a Function is One-to-One Using the Horizontal Line Test

Work through the video that accompanies Objective 2 and write notes here.

Write down the **Horizontal Line Test** and write down the 3 examples seen in the video.

Work through the animation that accompanies Example 1:

Example 1   **Determine Whether a Function Is One-to-One**
Determine whether each function is one-to-one.

a.

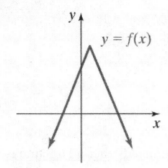

$y = f(x)$

b.

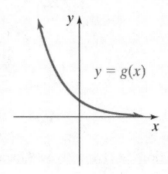

$y = g(x)$

c. $f(x) = x^2 + 1,\ x \le 0$ d. $f(x) = \begin{cases} 2x + 4 \text{ for } x \le -1 \\ 2x - 6 \text{ for } x \ge 4 \end{cases}$

Work parts c) and d). You will need to sketch these two functions.

c. $f(x) = x^2 + 1,\ x \le 0$ d. $f(x) = \begin{cases} 2x + 4 \text{ for } x \le -1 \\ 2x - 6 \text{ for } x \ge 4 \end{cases}$

Section 2.6 Objective 3 Understanding and Verifying Inverse Functions
Work through the video that accompanies Objective 3 and write notes here.

Write down the definition of an **inverse function.**

Write down the example of the one-to-one function and the inverse function given in this video.

Watch the video that follows Figure 27 and take notes here:

Write down the two **composition cancellation equations**:

Work through the interactive video that accompanies Example 2 and take notes here:

Show that $f(x) = \dfrac{x}{2x+3}$ and $g(x) = \dfrac{3x}{1-2x}$ are inverse functions using the composition cancellation equations.

Section 2.6 Objective 4 Sketching the Graphs of Inverse Functions
Read through Objective 4 and describe in your own words how to sketch the graph of the inverse of a given one-to-one function.

Below are the graphs of two one-to-one functions. Sketch the graphs of their inverse functions.

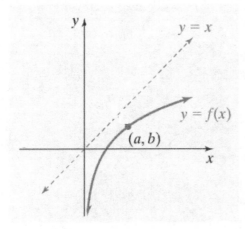

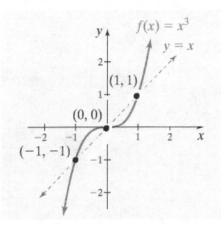

Work through the animation that accompanies Example 3:

Sketch the graph of $f(x) = x^2 + 1$, $x \le 0$, and its inverse. Also state the domain and range of f and f^{-1}.

Section 2.6 Objective 5 Finding the Inverse of a One-to-One Function

Work through the video that accompanies Objective 5 and take notes here:

In this video you are asked to find the inverse of $f(x) = x^2 + 1,\ x \le 0$. Write down the four steps for finding inverse functions and find the inverse of $f(x) = x^2 + 1,\ x \le 0$

Steps for Finding the Equation of an Inverse Function

      Step 1:

      Step 2:

      Step 3:

      Step 4:

97

Section 2.6

In your own words, explain the relationship between the domain and range of a one-to-one function and its inverse function:

Work through the animation that accompanies Example 4: Find the inverse function of $f(x) = \dfrac{2x}{1-5x}$ and state the domain and range of f and f^{-1}

Use the **Inverse Function Summary** to complete the following statements:

1. The function f^{-1} exists if and only if...

2. The domain of f is the same as the...

    And the range of f is the same as the...

3. To verify that two one-to-one functions, f and g, are inverses of each other, we must...

4. The graph of f^{-1} is a reflection of ...

    That is, for any point (a,b) that lies on the graph of f, the point (b,a) must...

5. To find the inverse of a one-to-one function, ...

Section 3.1 Guided Notebook

Section 3.1 Quadratic Functions

       Work through Section 3.1 TTK #1

       Work through Section 3.1 TTK #2

       Work through Section 3.1 TTK #3

       Work through Section 3.1 TTK #5

       Work through Objective 1

       Work through Objective 2

       Work through Objective 3

       Work through Objective 4

       Work through Objective 5

Section 3.1 Quadratic Functions

<u>3.1 Things To Know</u>

1. Solving Quadratic Equations by Factoring and the Zero Product Property

Can you solve the equation $5x^2 + 14x - 3 = 0$ by factoring? You should get an answer

of $x = -3$ or $x = \dfrac{1}{5}$. Try working through a "You Try It" problem, refer to Appendix B.4 or

watch the video.

2. Solving Quadratic Equations by Completing the Square

Do you remember how to complete the square? What number must be added to the binomial $x^2 + \dfrac{5}{6}x$ in order to complete the square? You should get an answer of $\dfrac{25}{144}$. Try working through a "You Try It" problem, refer to Appendix B.4 or watch the video.

3. Solving Quadratic Equations Using the Quadratic Formula

Can you solve the equation $2x^2 - 2x + 3 = 0$ using the quadratic formula? You should get an answer of $x = \dfrac{1 + i\sqrt{5}}{2}$ or $x = \dfrac{1 - i\sqrt{5}}{2}$. Try working through a "You Try It" problem or refer to Appendix B.4.

Section 3.1

5. Using Combinations of Transformations to Graph Functions
   Work through the animation and explain how to sketch the graph of
   $f(x) = -2(x+3)^2 - 1$.

Section 3.1 Objective 1 Understanding the Definition of a Quadratic Function and its Graph
Watch the video that accompanies Objective 1 and take notes here:

Write down the definition of a **quadratic function**:

Sketch two different quadratic functions. Sketch one quadratic function that "opens up" and sketch another quadratic function that "opens down".

What determines whether or not the graph of a quadratic function of the form $f(x) = ax^2 + bx + c$ opens up or down?

Work through Example 1 and take notes here: Without graphing, determine whether the graph of the quadratic function $f(x) = -3x^2 + 6x + 1$ opens up or down.

It is crucial that you understand the five basic characteristics of a parabola. Carefully work through the on page 3.1-5 and describe the following five characteristics of a parabola in your own words.

1. Vertex

2. Axis of Symmetry

3. y-intercept

4. x-intercept(s) or real zeros

5. Domain and range

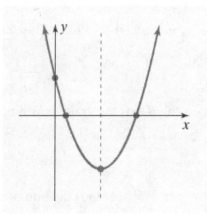

Section 3.1 Objective 2 Graphing Quadratic Functions Written in Standard Form

Section 3.1

Work through the animation and explain how to sketch the graph of $f(x) = -2(x+3)^2 - 1$.

Standard Form of a Quadratic Function (Fill in the Blanks)

A quadratic function is in **standard form** if it is written as

$f(x) =$ _____. The graph is a parabola with vertex _____.

The parabola "opens up" if _____. The parabola "opens down" if _____.

Work through the video that accompanies Example 2 and answer each of the following questions:

Given that the quadratic function $f(x) = -(x-2)^2 - 4$ is in standard form, address the following:

    a)   What are the coordinates of the vertex?

    b) Does the graph "open up" or "open down"?

    c) What is the equation of the axis of symmetry?

d) Find any x-intercepts.

e) Find the y-intercept.

f) Sketch the graph.

g) State the domain and range in interval notation.

Section 3.1 Objective 3 Graphing Quadratic Functions by Completing the Square
Work through the video that accompanies Example 3:

Rewrite the quadratic function $f(x) = 2x^2 - 4x - 3$ in standard form, and then answer
the questions below.

a) What are the coordinates of the vertex?

b) Does the graph "open up" or "open down"?

c) What is the equation of the axis of symmetry?

d) Find any x-intercepts.

e) Find the y-intercept.

f) Sketch the graph.

g) State the domain and range in interval notation.

Section 3.1 Objective 4 Graphing Quadratic Functions Using the Vertex Formula
Watch the video that accompanies Objective 4 and write your notes here:

Write down the formula for the vertex of the quadratic function $f(x) = ax^2 + bx + c$, $a \neq 0$.

Section 3.1

Work through Example 4:

Given the quadratic function $f(x) = -2x^2 - 4x + 5$, address the following:

a) What are the coordinates of the vertex?

b) Does the graph "open up" or "open down"?

c) What is the equation of the axis of symmetry?

d) Find any x-intercepts. (See page 3.1-15)

e) Find the y-intercept.

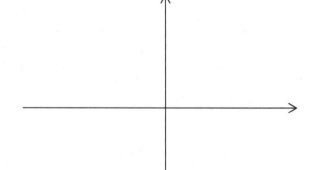

f) Sketch the graph.

g) State the domain and range in interval notation.

Section 3.1 Objective 5 Determining the Equation of a Quadratic Function Given Its Graph
Work through the video that accompanies Example 5: Analyze the graph to address the
following about the quadratic function it represents.

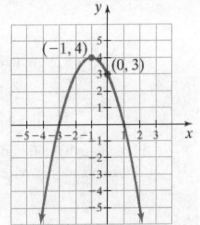

a) Is the leading coefficient positive or negative? (Why?)

b) What is the value of h? What is the value of k?

c) What is the value of the leading coefficient, a?

d) Write the equation of the function in standard form
$f(x) = a(x-h)^2 + k$.

e) Write the equation of the function in the form
$f(x) = ax^2 + bx + c$.

Section 3.2 Guided Notebook

Section 3.2 Applications and Modeling of Quadratic Functions
Work through Objective 1
Work through Objective 2
Work through Objective 3

Section 3.2 Applications and Modeling of Quadratic Functions

Work through the Introduction Video and take notes here: (Make sure that you write down the formula for the vertex of a quadratic function!)

Section 3.2 Objective 1 Maximizing Projectile Motion Functions

Work through Example 1: A toy rocket is launched with an initial velocity of 44.1 meters per second from a 1-meter tall platform. The height h of the object at any time t seconds after launch is given by the function $h(t) = -4.9t^2 + 44.1t + 1$. How long after launch did it take the rocket to reach its maximum height? What is the maximum height obtained by the toy rocket?

Work through Example 2: If an object is launched at an angle of 45 degrees from a 10-foot platform at 60 feet per second, it can be shown that the height of the object in feet is given by the quadratic function

$h(x) = -\dfrac{32x^2}{(60)^2} + x + 10$, where x is the horizontal distance of the object from the platform.

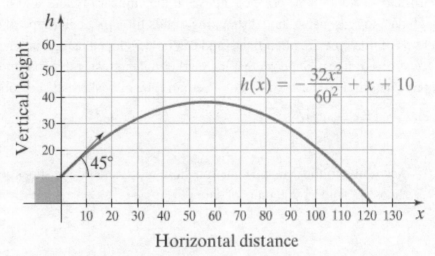

a)   What is the height of the object when its horizontal distance from the platform is 20 feet? Round to two decimal places.

b)   What is the horizontal distance from the platform when the object is at its maximum height?

c)   What is the maximum height of the object?

Section 3.2 Objective 2 Maximizing Functions in Economics
What is the equation that describes revenue?

Work through the video that accompanies Example 3:
Records can be kept on the price of shoes and the number of pairs sold in order to gather
enough data to reasonably model shopping trends for a particular type of shoe. Demand
functions of this type are often linear and can be developed using knowledge of the slope of
the equations of lines. Suppose that the marketing and research department of a shoe
company determined that the price of a certain type of basketball shoe obeys the demand
equation $p = -\dfrac{1}{50}x + 110$.

a) According to the demand equation, how much should the shoes sell for if 500 pairs of
    shoes are sold? 1,200 pairs of shoes?

b) What is the revenue if 500 pairs of shoes are sold? 1,200 pairs of shoes?

c) How many pairs of shoes should be sold in order to maximize revenue? What is the
    maximum revenue?

d) What price should be charged in order to maximize revenue?

Work through Example 4:

To sell x waterproof CD alarm clocks, WaterTime LLC has determined that the price in dollars must be $p = 250 - 2x$, which is the demand equation. Each clock costs $2 to produce, with fixed costs of $4,000 per month, producing the cost function of $C(x) = 2x + 4,000$.

a) Express the revenue R as a function of x.

b) Express the profit P as a function of x.

c) Find the value of x that maximizes profit. What is the maximum profit?

d) What is the price of the alarm clock that will maximize profit?

Work through Example 5:

A country club currently has 400 members who pay $500 per month for membership dues. The country club's board members want to increase monthly revenue by *lowering* the monthly dues in hopes of attracting new members. A market research study has shown that for each $1 decrease in monthly membership price, two additional people will join the club. What price should the club charge to maximize revenue? What is the maximum revenue?

Note that the video that accompanies Example 5 gives you the choice to see the solution using two different methods. Work through BOTH methods and decide for yourself which method is best for you.

Solve using the method shown in the text:

Now, solve the same problem using an alternate method as shown in the interactive video on the following page:

Example 5 using an **alternate method:**

A country club currently has 400 members who pay $500 per month for membership dues. The country club's board members want to increase monthly revenue by *lowering* the monthly dues in hopes of attracting new members. A market research study has shown that for each $1 decrease in monthly membership price, two additional people will join the club. What price should the club charge to maximize revenue? What is the maximum revenue?

Section 3.2

Section 3.2 Objective 3 Maximizing Area Functions

Suppose that you have 3,000 feet of fencing to construct the rectangular pen that borders a river as seen below. Watch the first video located at the top of p. 3.2-18 to see a couple of different ways that you can construct this fence. What should the length and width of this fence be in order to maximize area? What is the maximum area?

Work through the video that accompanies Example 6:

Mark has 100 feet of fencing available to build a rectangular pen for his hens and roosters. He wants to separate the hens and roosters by dividing the pen into two equal areas. What should the length of the center partition be in order to maximize area? What is the maximum area?

Section 3.3 Guided Notebook

Section 3.3 The Graphs of Polynomial Functions
    Work through Objective 1
    Work through Objective 2
    Work through Objective 3
    Work through Objective 4
    Work through Objective 5
    Work through Objective 6
    Work through Objective 7

Section 3.3 The Graphs of Polynomial Functions

Section 3.3 Objective 1 Understanding the Definition of a Polynomial Function
Work through the video that accompanies Objective 1 and take notes here:

What is the definition of a **polynomial function**?

Determine which functions are polynomial functions. If the function is a polynomial function, identify the degree, the leading coefficient, and the constant coefficient.

$$f(x) = \sqrt{3}x^3 - 2x^2 - \frac{1}{6}$$

$$g(x) = 4x^5 - 3x^3 + x^2 + \frac{7}{x} - 3$$

Now work through Example 1 (Part c): Determine if the function is a polynomial function. If the function is a polynomial function, identify the degree, the leading coefficient, and the constant coefficient.

$$h(x) = \frac{3x - x^2 + 7x^4}{9}$$

Section 3.3 Objective 2 Sketching the Graphs of Power Functions
The graphs of five power functions are seen here.

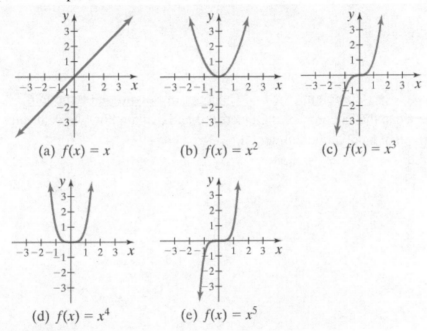

(a) $f(x) = x$ (b) $f(x) = x^2$ (c) $f(x) = x^3$

(d) $f(x) = x^4$ (e) $f(x) = x^5$

Work through the interactive video that accompanies Example 2 and use a power function to sketch the following functions:

a) $f(x) = -x^6$

    b) $f(x) = (x+1)^5 + 2$

    c) $f(x) = 2(x-3)^4$

Section 3.3 Objective 3 Determining the End Behavior of Polynomial Functions

Work through the video that accompanies Objective 3 and explain how we determine the end behavior of the graph of a polynomial function.

Write down the Two-Step Process for Determining the End Behavior of a Polynomial Function as seen on pages 3.3-11 through 3.3-13.

Step 1:

Step 2:

Work through the video that accompanies Example 3: Use the end behavior of each graph to determine whether the degree is even or odd and whether the leading coefficient is positive or negative.

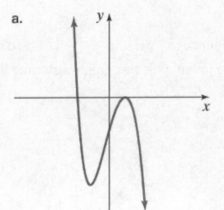

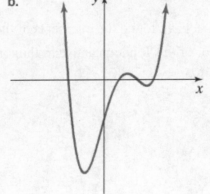

121

Section 3.3

<u>Section 3.3 Objective 4 Determining the Intercepts of a Polynomial Function</u>
Read page 3.3-15 and answer the following:
Explain how to determine the *y*-intercept of a polynomial function:

Explain how to find the *x*-intercepts (also called **real zeros**) of a polynomial function:

Work through Example 4: Find the intercepts of the polynomial function
$f(x) = x^3 - x^2 - 4x + 4$.

Now that you have found the intercepts of the function $f(x) = x^3 - x^2 - 4x + 4$, determine the end behavior of this function and start thinking about what this graph is starting to look like. (See p. 3.3-17)

Section 3.3 Objective 5 Determining the Real Zeros of Polynomial Functions and Their Multiplicities

Work through the video that accompanies Objective 5 and take notes and fill in all blanks below:

If $(x-c)$ is a factor of a polynomial function then _____.

(fill in the blank)

Multiplicity

If $(x-c)^k$ is a factor of a polynomial function where k is a positive integer then

_____ .

(fill in the blank)

If $(x-c)^k$ is a factor of a polynomial function where c is a real number and k is an integer such that $k \geq 1$, then

- If k is odd, then _____

- If k is even, then_____

The last part of the Objective 5 video (from page 3.3-18) shows the solution to Example 5:

Find all real zeros of $f(x) = x(x^2 - 1)(x - 1)$. Determine the multiplicities of each zero, and decide whether the graph touches or crosses at each zero. (Try to create a rough sketch of this graph.)

Section 3.3 Objective 6 Sketching the Graph of a Polynomial Function
Carefully work through pages 3.3-22 – 3.3-25 and answer the following questions:

Can you rewrite the function $f(x) = x^3 - x^2 - 4x + 4$ in factored form? What is the factored form of this function?

What are the three zeros and their multiplicities of the function $f(x) = x^3 - x^2 - 4x + 4$? What is the y-intercept of this function?

Now plot the three zeros and plot the y-intercept also use the end behavior to begin to create a rough sketch of the function. (See Figure 16 on page 3.3-22)

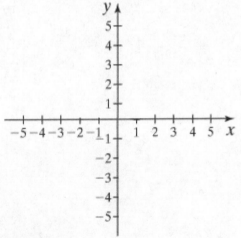

What is the definition of a **test value**?

Choose two test values as shown on p. 3.3-23 then complete a "rough sketch" of $f(x) = x^3 - x^2 - 4x + 4$ on the grid above.

What is the definition of a **turning point**?

125

How many turning points can the graph of a polynomial of degree n have?

What is a turning point in which the graph changes from increasing to decreasing called?

What is a turning point in which the graph changes from decreasing to increasing called?

Write down the **Four-Step Process for Sketching the Graph of Polynomial Functions.**

        Step 1:

        Step 2:

        Step 3:

        Step 4:

Work through the interactive video that accompanies Example 6:

Use the four-step process to sketch the graphs of the following polynomial functions.

a) $f(x) = -2(x+2)^2(x-1)$

b) $f(x) = x^4 - 2x^3 - 3x^2$

Section 3.3 Objective 7 Determining a Possible Equation of a Polynomial Function Given Its Graph

Work through the Interactive Video that accompanies Example 7:

Analyze the graph to address the following about the polynomial function it represents.

a) Is the degree of the polynomial function even or odd?

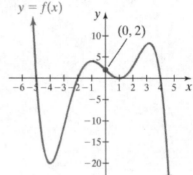

b) Is the leading coefficient positive or negative?

c) What is the value of the constant coefficient?

d) Identify the real zeros, and state the multiplicity of each.

e) Select from this list a possible function that could be represented by this graph.

   i. $f(x) = -\dfrac{1}{20}(x+5)(x+2)(x-1)^2(x-4)$

   ii. $f(x) = -\dfrac{1}{800}(x+5)^2(x+2)^2(x-1)(x-4)^2$

   iii. $f(x) = \dfrac{1}{20}(x+5)(x+2)(x-1)^2(x-4)$

   iv. $f(x) = -\dfrac{1}{10}(x+5)(x+2)(x-1)^2(x-4)$

Section 3.4 Guided Notebook

Section 3.4 Synthetic Division; The Remainder and Factor Theorems

Work through Section 4.1 TTK #1

Work through Section 4.1 TTK #3

Work through Section 4.1 TTK #5

Work through Objective 1

Work through Objective 2

Work through Objective 3

Work through Objective 4

Work through Objective 5

Section 3.4 Synthetic Division; The Remainder and Factor Theorems

3.4 Things To Know

1. Using Long Division to Divide Polynomials

Watch the video on p. 3.4-1 to see how to find the quotient when $3x^4 + x^3 + 7x + 4$ is divided by $x^2 - 1$.

3. Determining the Intercepts of a Function

Do you remember how to find intercepts of a function? Try this one: Find the intercepts of $f(x) = x^3 + 6x^2 - x - 6$. You should get an answer of $-6, -1,$ and 1 for the x-intercepts and -6 for the y-intercept. Try working through a "You Try It" problem or refer to Section 2.2 or watch the video.

5. Sketching the Graph of a Polynomial Function

Work through the interactive video to recall how to use the Four-Step Process for Sketching Polynomial Functions. Work through the video to sketch

a) $f(x) = -2(x+2)^2(x-1)$ and b) $f(x) = x^4 - 2x^3 - 3x^2$.

Section 3.4 Objective 1 Using the Division Algorithm

Write down the **Division Algorithm**

Work through Example 1: Given the polynomials $f(x) = 4x^4 - 3x^3 + 5x - 6$ and $d(x) = x - 2$, find polynomials $q(x)$ and $r(x)$, and write $f(x)$ in the form $f(x) = d(x)q(x) + r(x)$.

Write down the **Corollary to the Division Algorithm**

131

Section 3.4

Section 3.4

Section 3.4 Objective 2 Using Synthetic Division

Synthetic division is a very handy way to long divide polynomials when the divisor is of the form $x - c$. Carefully work through the video that accompanies Objective 2 and **use synthetic division** to divide $f(x) = 4x^4 - 3x^3 + 5x - 6$ by $x - 2$.

Work through the video that accompanies Example 2: Use synthetic division to divide $f(x)$ by $x - c$, and then write $f(x)$ in the form $f(x) = (x - c)q(x) + r$ for

$f(x) = -2x^4 + 3x^3 + 7x^2 - x + 5$ divided by $x + 1$.

132

Section 3.4 Objective 3 Using the Remainder Theorem

Carefully read page 3.4-11 and take notes here:

Write down **The Remainder Theorem**

Work through the video that accompanies Example 3:

Use the remainder theorem to find the remainder when $f(x)$ is divided by $x-c$.

a) $f(x) = 5x^4 - 8x^2 + 3x - 1; \ x - 2$

b) $f(x) = 3x^3 + 5x^2 - 5x - 6; \ x + 2$

Section 3.4

Section 3.4 Objective 4 Using the Factor Theorem

What is the definition of a **factor**?

What does it mean for a polynomial to be a factor of another polynomial?

Write down the **Factor Theorem.**

Work through the video that accompanies Example 4:

Determine whether $x+3$ is a factor of $f(x) = 2x^3 + 7x^2 + 2x - 3$.

Section 3.4 Objective 5 Sketching the Graph of a Polynomial Function
Carefully read page 3.4-15 and show how to completely factor the polynomial
$f(x) = 2x^3 + 7x^2 + 2x - 3$.

$f(x) = $ _____

(Write $f(x) = 2x^3 + 7x^2 + 2x - 3$ in completely factored form)

Once you have shown how to completely factor $f(x) = 2x^3 + 7x^2 + 2x - 3$, watch the video that can be found in the last sentence on page 3.4-15 and use the four-step process to sketch the graph of $f(x) = 2x^3 + 7x^2 + 2x - 3$.

Work through the video that accompanies Example 5: Given that $x = 2$ is a zero of $f(x) = x^3 - 6x + 4$, completely factor f and sketch its graph. (Be sure to write f in completely factored form.)

Section 3.5 Guided Notebook

Section 3.5 The Zeros of Polynomial Functions; The Fundamental Theorem of Algebra

> Work through TTK # 1
> Work through TTK # 2
> Work through TTK # 3
> Work through TTK # 4
>
> Work through Objective 1
> Work through Objective 2
> Work through Objective 3
> Work through Objective 4
> Work through Objective 5
> Work through Objective 6

Section 3.5 The Zeros of Polynomial Functions;

The Fundamental Theorem of Algebra

Section 3.5 Objective 1 Using the Rational Zeros Theorem

Look at your previous page of notes and see that we were given that $x = 2$ was a zero of $f(x) = x^3 - 6x + 4$. We used this information to completely factor f as $f(x) = (x-2)\left(x-(-1+\sqrt{3})\right)\left(x-(-1-\sqrt{3})\right)$. What if we were not given the fact that $x = 2$ was a zero? How would we have found the three zeros? The answer is that we first must use the **rational zeros theorem.**

What is the definition of a **rational zero**?

Section 3.5

Write down the **Rational Zeros Theorem.**

Work through Example 1: Use the rational zeros theorem to determine the potential zeros of the polynomial function $f(x) = 4x^4 - 7x^3 + 9x^2 - x - 10$.

Section 3.5 Objective 2 Finding the Zeros of a Polynomial Function
Carefully read pages 3.5-5 and 3.5-6.

Write down the **Fundamental Theorem of Algebra.**

Write down the **Number of Zeros Theorem.**

Work through the video that accompanies Example 2: Find all complex zeros of $f(x) = 6x^4 + 13x^3 + 61x^2 + 8x - 10$ and rewrite $f(x)$ in completely factored form.

Note: Once we factor a polynomial into the product of linear factors and a quadratic function of the form $f(x) = (x - c_1)(x - c_2)(x - c_3) \cdots (x - c_k)(ax^2 + bx + c)$, we need simply to solve the quadratic equation $ax^2 + bx + c = 0$ to find the remaining two zeros. For example, if we want to find all zeros of $g(x) = 2x^3 - 3x^2 + 4x - 3$ we need only try to find **one** zero from the list created by the Rational Zeros Theorem then find the two zeros of the remaining quadratic function.

See if you can find the zeros of $g(x) = 2x^3 - 3x^2 + 4x - 3$. Watch the **video** by clicking on the word "video" seen on the bottom of page 3.5-9.

Section 3.5 Objective 3 Solving Polynomial Equations
Work through the video that accompanies Example 3:

$$\text{Solve } 5x^5 - 9x^4 + 23x^3 - 35x^2 + 12x + 4 = 0$$

Section 3.5 Objective 4 Using the Complex Conjugate Pairs Theorem
Read page 3.5-12.

Write down the **Complex Conjugate Pairs Theorem**

Work through the interactive video that accompanies Example 4:

Find the equation of a polynomial function f with real coefficients that satisfies the given conditions:

a) Fourth-degree polynomial function such that 1 is a zero of multiplicity 2 and $2-i$ is also a zero.

Continued on next page

Example 4 b) Find the fifth-degree polynomial function sketched below given that $1+3i$ is a zero.

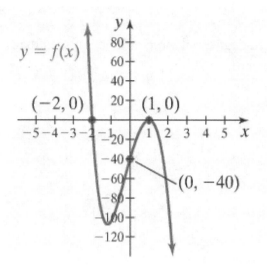

Section 3.5 Objective 5 Using the Intermediate Value Theorem

Why is it true that every odd degree polynomial with real coefficients has at least one real zero? The answer to this question is on page 3.5-16.

Write down the **Intermediate Value Theorem** and sketch the graph seen on page 3.5-17.

Work through the video that accompanies Example 5:

Use the Intermediate Value Theorem to find the real zero of $f(x) = x^3 + 2x - 1$ correct to two decimal places.

143

Section 3.5

Section 3.5 Objective 6 Sketching the Graphs of Polynomial Functions
Write down the **Steps for Sketching the Graphs of Polynomial Functions** as seen on page 3.5-20:

    Step 1

    Step 2

    Step 3

    Step 4

Work through the video that accompanies Example 6:

    Sketch the graph of $f(x) = 2x^5 - 5x^4 - 2x^3 + 7x^2 - 4x + 12$.

Section 3.6 Guided Notebook

Section 3.6 Rational Functions and Their Graphs

- ☐ Work through TTK# 1
- ☐ Work through TTK# 2
- ☐ Work through TTK# 3
- ☐ Work through Objective 1
- ☐ Work through Objective 2
- ☐ Work through Objective 3
- ☐ Work through Objective 4
- ☐ Work through Objective 5
- ☐ Work through Objective 6
- ☐ Work through Objective 7

Section 3.6 Rational Functions and Their Graphs

Work through the interactive video for You Try It TTK #1 **Determining the Domain of a Function Given the Equation** and take notes here.

Work through the video for You Try It TTK #2 **Determining Whether a Function is Even, Odd, or Neither** and take notes here.

145

Section 3.6

Work through the animation for You Try It TTK #3 **Using Combinations of Transformations to Graph Functions** and take notes here.

Section 3.6 Objective 1: Finding the Domain and Intercepts of Rational Functions

Write down the definition of a **Rational Function**.

Where are rational functions not defined?

How do you find the y-intercept if there is one?

How do you find the x-intercepts?

Work Example 1 showing all work below. To see the steps of the solution watch the video with the example.

Let $f(x) = \dfrac{x-4}{x^2+x-6}$.

a.   Determine the domain of f.

b.   Determine the y-intercept (if any)

c.   Determine any x-intercepts.

Section 3.6

Section 3.6 Objective 2: Identifying Vertical Asymptotes

What is a **vertical asymptote** and when does it occur?

What is the caution statement say about locating vertical asymptotes?

Watch the video with Example 2 and answer the following questions.

Find the vertical asymptotes (if any) of the function $f(x) = \dfrac{x-3}{x^2+x-6}$ and then sketch the graph near the vertical asymptotes.

1. What is the first step?

2. What are the vertical asymptotes?

3. When graphing vertical asymptotes the lines must be _____?

4. How do you sketch the graph near the asymptotes?

5. Show the work for each of the test values below. Include a sketch.

6. What is the side note about choosing the values and the x intercept?

Work through the video with Example 3 and take notes below.

Find the vertical asymptotes (if any) of the following function and then sketch the graph near the vertical asymptotes.

$$f(x) = \frac{x+3}{x^2+x-6}$$

1. What is the first step?

2. What happens to the common factor?

3. What is the vertical asymptote?

4. Show all steps to sketch the graph near the asymptote.

Section 3.6 Objective 3: Identifying Horizontal Asymptotes

Watch the video with Objective 3 and answer the following questions.

1. Draw the sketch of the reciprocal function.

2. What do you notice about the graph and the y axis?

3. What do you notice about the graph and the x axis?

4. What is the x axis?

5. Write the definition of a horizontal asymptote.

6. Draw the illustration for the first 2 properties of Horizontal asymptotes of Rational Functions.

Write down the Definition of Horizontal Asymptotes.

Section 3.6

What are the 3 properties of Horizontal Asymptotes?

What is the technique for finding horizontal asymptotes of a rational function?

Work through Example 4 and show all work below. For detailed step by step solutions watch the video.

Find the horizontal asymptote of the graph of each rational function or state that one does not exist.

a) $f(x) = \dfrac{x}{x^2 - 4}$ b) $f(x) = \dfrac{4x^2 - x + 1}{1 - 2x^2}$ c) $f(x) = \dfrac{2x^3 + 3x^2 - 2x - 2}{x - 1}$

Section 3.6 Objective 4: Using Transformations to Sketch the graphs of Rational Functions.

Draw the graphs and write the properties of $f(x) = \dfrac{1}{x}$ and $f(x) = \dfrac{1}{x^2}$.

Work through Example 5 showing all work below. The video will show the step by step solution.

Use transformations to sketch the graph of $f(x) = \dfrac{-2}{(x+3)^2} + 1$.

Section 3.6

Section 3.6 Objective 5: Sketching Rational Functions Having Removable Discontinuities.

What are removable discontinuities? When do they occur?

Work through Example 6 showing all work below. For a detailed step by step solution watch the video.

Sketch the graph of the following function and find the coordinates of all removable discontinuities. $f(x) = \dfrac{x^2 - 1}{x + 1}$.

Work through Example 7 showing all work below. For a detailed step by step solution watch the video with the example.

Sketch the graph of the following function and find the coordinates of all removable discontinuities. $f(x) = \dfrac{x+3}{x^2+x-6}$.

What is the caution statement about graphing calculators and removable discontinuities?

Section 3.6 Objective 6: Identifying Slant Asymptotes.

When does a slant asymptote occur?

Section 3.6

How do you find the slant asymptote?

Watch the video with Example 8 and take notes here.

Find the slant asymptote of $f(x) = \dfrac{2x^2 + 3x - 2}{x - 1}$.

Section 3.6 Objective 7: Sketching Rational Functions

What are the steps for Graphing Rational Functions of the Form $f(x) = \dfrac{g(x)}{h(x)}$?

Section 3.6

Watch the video with Example 9 answering the following questions.

$$\text{Sketch the graph of} \quad f(x) = \frac{x^3 + 2x^2 - 9x - 18}{x^3 + 6x^2 + 5x - 12}.$$

1. Write down the steps to find the domain.

2. After finding the domain, what is the next step? Show all work for this step below.

3. What is the simplified function?

4. What happens at x = -3?

5. Is the function even or odd? What does that indicate about the graph?

6. Why is it a good idea to test for symmetry?

7. Show the steps to find the intercepts below.

8. Which denominator is used to find the vertical asymptotes? What are they?

9. Verify the behavior near the vertical asymptotes below.

10. Will there be a horizontal or slant asymptote? Why?

11. What is the horizontal or slant asymptote? How was it found?

12. How do you determine which values of x to evaluate?

13. Show the steps to evaluate the function at these values.

14. Draw the final sketch of the graph below.

Section 3.7 Guided Notebook

Section 3.7 Variation

Work through TTK #1

Work through Objective 1

Work through Objective 2

Work through Objective 3

Section 3.7 Variation

Work through the You Try It Exercise: **Converting Verbal Statements into Mathematical Statements** showing all steps below.

Section 3.7 Objective 1: Solve Application Problems involving Direct Variation.

What is the purpose of variation equations?

What is **Direct Variation**?

What is the **constant of variation**?

Section 3.7

What are the four steps for **Solving Variation Problems**?

Write down all the steps for solving Example 1.

> The kinetic energy of an object in motion varies directly with the square of its speed. If a van traveling at a speed of 30 meters per second has 945,000 joules of kinetic energy, how much kinetic energy does it have if it is traveling at a speed of 20 meters per second?

Work through Example 2 showing all steps below.

> The Ponderal Index measure of leanness states that body mass varies directly with the cube of height. If a "normal" person who is 1.2 m tall has a body mass of 21.6 kg, then what is the body mass of a "normal" person that is 1.8 m tall?

Section 3.7 Objective 2: Solve Application Problems Involving Inverse Variation

What is **Inverse Variation**?

Show all the steps for Example 3 below.

For a given mass, the density of an object is inversely proportional to its volume. If 50 cubic centimeters of an object with a density of $28\dfrac{g}{cm^3}$ is compressed to 40 cubic centimeters, what would be its new density?

Work through Example 4 showing all steps below.

The shutter speed, S, of a camera varies inversely as the square of the aperture setting, f. If the shutter speed is 125 for an aperture of 5.6, what is the shutter speed if the aperture is 1.4?

Section 3.7 Objective 3: Solve Application Problems Involving Combined Variation.

What is **combined variation**? Write down 3 examples.

In general, when are variables directly related? Inversely related?

What is **joint variation**?

Show all steps for Example 5 below.

The number of gallons of a liquid that can be stored in a conical tank is directly proportional to the area of the base of the tank and its height (joint variation). A tank with a base area of 1200 square feet and a height of 15 feet holds 45,000 gallons of liquid. How tall must the tank be to hold 75,000 gallons of liquid if its base area is 1500 square feet?

Show all the steps for Example 6 below.

> The resistance of a wire varies directly with the length of the wire and inversely with the square of its radius. A wire with a length of 500 cm and a radius of 0.5 cm has a resistance of 15 ohms. Determine the resistance of an 800 cm piece of similar wire with a radius of 0.8 cm.

Work through Example 7 showing all steps below.

> For a fixed speed, the number of calories burned while jogging varies jointly with the weight of the jogger (in kg) and the time spent jogging (in minutes). If a 100-kg man jogs for 40 minutes and burns 490 calories, how many calories will a 130-kg man burn if he jogs for 60 minutes at the same speed?

Section 4.1 Guided Notebook

Section 4.1 Exponential Functions
Work through Section 4.1 TTK #1
Work through Section 4.1 TTK #2
Work through Objective 1
Work through Objective 2
Work through Objective 3
Work through Objective 4

Section 4.1 Exponential Functions

4.1 Things To Know

1. Using Combinations of Transformations to Graph Functions

Do you remember how to sketch functions using transformations? You will need to use transformations to sketch exponential functions in this section. Work through the **animation** to refresh your memory. Write down the "Order of Transformations" as seen in this animation.

2. Determining Whether a Function is One-to-One Using the Horizontal Line Test
You must understand the **definition of a one-to-one** function. Go back to Section 2.6 to see this definition or work through the video. You should also work through the animation to recall how to use the horizontal line test.

Section 4.1 Objective 1 Understanding the Characteristics of Exponential Functions

Read page 4.1-2. Write down the **definition of an exponential function.**

169

Section 4.1

Watch the video seen on page 4.1-2 and take notes here: **WRITE DOWN THE CHARACTERISTICS OF THE EXPONENTIAL FUNCTION AS SEEN NEAR THE END OF THIS VIDEO.**

Work through the video that accompanies Example 1 and take notes here:

Sketch the graph of $f(x) = \left(\dfrac{2}{3}\right)^x$.

Work through Example 2 and take notes here:

Find the exponential function $f(x) = b^x$ whose graph is given as follows.

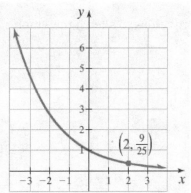

Section 4.1

Section 4.1 Objective 2 Sketching the Graphs of Exponential Functions Using Transformations

Read page 4.1-10. Sketch the graphs of $y = 3^x$ and $y = 3^x - 1$ here:

Work through the video that accompanies Example 3 and take notes here:

Use transformations to sketch the graph of $f(x) = -2^{x+1} + 3$.

<u>Section 4.1 Objective 3 Solving Exponential Equations by Relating the Bases</u>

Why is every exponential function of the form $f(x) = b^x$ one-to-one?

A function f is one-to-one if for any two range values $f(u)$ and $f(v)$,

$f(u) = f(v)$ implies that _____.

(Fill in the Blank)

Method of Relating the Bases

If an exponential function can be written in the form $b^u = b^v$, then

_____.

(Fill in the Blank)

Carefully work through the **animation** that accompanies Example 4 and take notes here:

Solve the following equations:

a) $8 = \dfrac{1}{16^x}$

b) $\dfrac{1}{27^x} = \left(\sqrt[4]{3}\right)^{x-2}$

Section 4.1

Section 4.1 Objective 4 Solving Applications of Exponential Functions

Work through Example 5:

> Most golfers find that their golf skills improve dramatically at first and then level off rather quickly. For example, suppose that the distance (in yards) that a typical beginning golfer can hit a 3-wood after t weeks of practice on the driving range is given by the exponential function $d(t) = 225 - 100(2.7)^{-0.7t}$. This function has been developed after many years of gathering data on beginning golfers.

> How far can a typical beginning golfer initially hit a 3-wood? How far can a typical beginning golfer hit a 3-wood after 1 week of practice on the driving range? After 5 weeks? After 9 weeks? Round to the nearest hundredth yard.

Compound Interest

What is simple interest?

Read pages 4.1-17 and 4.1-18 to see how the formula for periodic compound interest is derived.

Write the **formula for periodic compound interest** here:

For interest compounded…
annually, use $n =$ _____
semi-annually, use $n =$ _____
quarterly, use $n =$ _____
monthly, use $n =$ _____
daily, use $n =$ _____

Work through Example 6:

Which investment will yield the most money after 25 years?

Investment A: $12,000 invested at 3% compounded monthly.
Investment B: $10,000 invested at 3.9% compounded quarterly.

175

Present Value

Present value is the amount of money needed now (in the present) to reach an investment goal in the future.

Show how to derive the formula for present value by starting with the formula for periodic compound interest here:

Write down the **formula for present value** here:

Work through Example 7:

> Find the present value of $8000 if interest is paid at a rate of 5.6% compounded quarterly for 7 years. Round to the nearest cent.

Section 4.2 Guided Notebook

Section 4.2 The Natural Exponential Function
Work through Objective 1
Work through Objective 2
Work through Objective 3
Work through Objective 4

Section 4.2 The Natural Exponential Function

Section 4.2 Objective 1 Understanding the Characteristics of the Natural Exponential Functions
Watch the video that accompanies Objective 1 and takes notes here:

What is the number *e* rounded to 6 decimal places?

Sketch the graphs of $y = 2^x$, $y = 3^x$, and $y = e^x$ on the same grid.

Make sure that you have a scientific calculator that has an $\boxed{e^x}$ key. Then do Example 1:

Evaluate each expression correctly to six decimal places.

a) e^2 b) $e^{-.534}$ c) $1000e^{.013}$

Section 4.2 Objective 2 Sketching the Graphs of Natural Exponential Functions Using Transformations

Work through the video that accompanies Example 2:

Use transformations to sketch the graph of $f(x) = -e^x + 2$.

Section 4.2 Objective 3 Solving Natural Exponential Equations by Relating the Bases

Write down the **method of relating the bases:**

Section 4.2

Work through the interactive video that accompanies Example 3:

Use the method of relating the bases to solve each exponential equation.

a) $e^{3x-1} = \dfrac{1}{\sqrt{e}}$

b) $\dfrac{e^{x^2}}{e^{10}} = \left(e^x\right)^3$

Section 4.2 Objective 4 Solving Applications of the Natural Exponential Function

Continuous Compound Interest

Watch the animation on page 4.2-8 to see how to derive the formula for continuous compound interest.

Write the **formula for continuous compound interest** here:

Work through Example 4:

How much money would be in an account after 5 years if an original investment of $6,000 was compounded continuously at 4.5%? Compare this amount to the same investment that was compounded daily.

Section 4.2

Present Value
Write the **formula for the present value** of money that is compounded continuously.

Work through Example 5:

Find the present value of $18,000 if interest is paid at a rate of 8% compounded continuously for 20 years. Round to the nearest cent.

Exponential Growth Model
Write down the exponential growth model as shown on page 4.2-12.

182

Work through the video that accompanies Example 6:

The population of a small town follows the exponential growth model $P(t) = 900e^{.015t}$, where t is the number of years after 1900. Answer the following questions, rounding each answer to the nearest whole number.

    a. What was the population of this town in 1900?

    b. What was the population of this town in 1950?

    c. Use this model to predict the population of this town in 2012.

Work through the video that accompanies Example 7:

Twenty years ago, the State of Idaho Fish and Game Department introduced a new breed of wolf into a certain Idaho forest. The current wolf population in this forest is now estimated at 825, with a relative growth rate of 12%. Answer the following questions, rounding each answer to the nearest whole number.

    a. How many wolves did the Idaho Fish and Game Department introduce into this forest?

    b. How many wolves can be expected after another 20 years?

Section 4.3 Logarithmic Functions

Work through Section 4.3 TTK #1

Work through Section 4.3 TTK #2

Work through Section 4.3 TTK #3

Work through Section 4.3 TTK #4

Work through Section 4.3 TTK #5

Work through Section 4.3 TTK #6

Work through Section 4.3 TTK #7

Work through Section 4.3 TTK #8

Work through Section 4.3 TTK #9

Work through Objective 1

Work through Objective 2

Work through Objective 3

Work through Objective 4

Work through Objective 5

Work through Objective 6

Work through Objective 7

Section 4.3 Logarithmic Functions

4.3 Things To Know

1. Solving Polynomial Inequalities (Appendix B.9)

   Review this objective as necessary. You will have to be able to solve a polynomial inequality later on during this homework assignment. Can you solve the inequality $2x^2 + 5x - 3 \geq 0$? You should get a solution of $(-\infty, -3] \cup [\frac{1}{2}, \infty)$. Watch the appropriate video and review Appendix B.9 as needed.

2. Solving Rational Inequalities (Appendix B.9)

   Review this objective as necessary. You will have to be able to solve a rational inequality later on during this homework assignment. Can you solve the inequality $\dfrac{x^2 - 3x - 10}{x + 4} \geq 0$? You should get a solution of $(-4, -2] \cup [5, \infty)$. Watch the appropriate video and review Appendix B.9 as needed.

3. Determining Whether a Function is One-to-One Using the Horizontal Line Test

   Do you remember the **definition of a one-to-one** function? Go back to Section 2.6 to see this definition or work through the video. You should also work through the animation to recall how to use the horizontal line test. Pay close attention to the graph in part (b) of this animation. Does the graph in part (b) of the animation represent a one-to-one function?

4. Sketching the Graph of an Inverse Function

Work through the animation and sketch the graph of the function $f(x) = x^2 + 1$, $x \leq 0$ and sketch the graph of its inverse function. Also, state the domain and range of f and f^{-1}.

5. Using the Composition Cancellation Properties of Inverse Functions

Refer back to Section 2.6 or work through the interactive video and write down the two composition cancellation equations.

Cancellation Equation 1:

Cancellation Equation 2:

6. Finding the Equation of an Inverse Function

   Do you remember how to find the inverse of a one-to-one function? Work through the animation to find the inverse of $f(x) = \dfrac{2x}{1-5x}$ and state the domain and range of f and f^{-1}.

7. Sketching the Graphs of Exponential Functions

   Sketch the graphs of $y = 2^x$ and $y = \left(\dfrac{1}{2}\right)^x$ as seen in the video.

187

8. Sketching the Graphs of Exponential Functions Using Transformations

   Sketch the graphs of $y = 2^x$ and $f(x) = -2^{x+1} + 3$ as seen in the video.

9. Solving Exponential Equations by Relating the Bases

   Can you solve these two equations? a) $8 = \dfrac{1}{16^x}$ b) $\dfrac{1}{27^x} = \left(\sqrt[4]{3}\right)^{x-2}$

   Work through the animation to see the solutions.

   a) $8 = \dfrac{1}{16^x}$

   b) $\dfrac{1}{27^x} = \left(\sqrt[4]{3}\right)^{x-2}$

Section 4.3 Objective 1 Understanding the Definition of a Logarithmic Function

Work through the video that accompanies Objective 1 and take notes here:

Sketch the graph of $f(x) = b^x$, $b > 1$ as seen in the video and plot several points that lie on the graph, then sketch the graph of the inverse function.

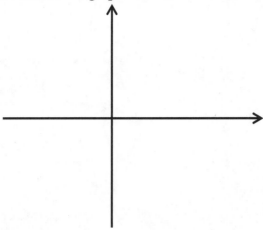

We want to find the equation of the inverse of $f(x) = b^x$. Follow the 4-step process for finding inverse functions as stated in this video and see if we can find the equation of this inverse function.

Step 1:

Step 2:

Step 3:

Before we can complete Step 3, we must define the logarithmic function. Write the definition of a logarithmic function here:

Definition of the Logarithmic Function

Now Finish Step 3:

Step 4:

Section 4.3

Work through the video that accompanies Example 1:

Write each exponential equation as an equation involving a logarithm.

a) $2^3 = 8$ b) $5^{-2} = \dfrac{1}{25}$ c) $1.1^M = z$

Work through the video that accompanies Example 2:

Write each exponential equation as an equation involving an exponent.

a) $\log_3 81 = 4$ b) $\log_4 16 = y$ c) $\log_{3/5} x = 2$

Section 4.3 Objective 2 Evaluating Logarithmic Expressions
Read through page 4.3-7 and take notes here.

Explain how to evaluate the expression $\log_4 64$

Write down "The Method of Relating the Bases".

Work through the interactive video that accompanies Example 3 and take notes here:
Evaluate each logarithm:

a) $\log_5 25$ b) $\log_3 \dfrac{1}{27}$ c) $\log_{\sqrt{2}} \dfrac{1}{4}$

191

Section 4.3 Objective 3 Understanding the Properties of Logarithms

Write down the two General Properties of Logarithms seen on page 4.3-9.

General Properties of Logarithms

For $b > 0$ and $b \neq 1$,

1.

2.

In Section 2.6, we studied one-to-one functions. Given a function f and the inverse function f^{-1}, we learned two composition cancellation equations. Write down the two composition cancellation equations here:

Composition Cancellation Equations

1.

2.

If $f(x) = b^x$ is an exponential function, the inverse function is $f^{-1}(x) = \log_b x$. Using this information and applying the two composition cancellation equations seen above, write down the two cancellation properties of exponentials and logarithms.

Cancellation Properties of Exponentials and Logarithms

For $b > 0$ and $b \neq 1$,

1.

2.

Work through Example 4:

Use the properties of logarithms that you have written above to evaluate each expression.

a) $\log_3 3^4$ b) $\log_{12} 12$ c) $7^{\log_7 13}$ d) $\log_8 1$

Section 4.3 Objective 4 Using the Common and Natural Logarithms
Read page 4.3-11 and take notes here. Then write down the definition of the **common logarithmic function** and the **natural logarithmic function.**

Common Logarithmic Function

Natural Logarithmic Function

Work through the video that accompanies Example 5:

Write each exponential equation as an equation involving a common logarithm or natural logarithm.

a) $e^0 = 1$

b) $10^{-2} = \dfrac{1}{100}$

c) $e^K = w$

Section 4.3

Work through the video that accompanies Example 6:

Write each logarithmic equation as an equation involving an exponent.

a) $\log 10 = 1$ b) $\ln 20 = Z$ c) $\log(x-1) = T$

Work through the video that accompanies Example 7:

Evaluate each expression without the use of a calculator.

a) $\log 100$ b) $\ln \sqrt{e}$ c) $e^{\ln 51}$ d) $\log 1$

Section 4.3 Objective 5 Understanding the Characteristics of Logarithmic Functions
Write down the three steps to sketching the graph of a logarithmic function of the form
$f(x) = \log_b x, \; b > 0, \; b \neq 1$ as seen on page 4.3-15.

Step 1:

Step 2:

Step 3:

Follow these 3 steps to sketch the graph of $f(x) = \log_3 x$. Work through the video that
accompanies Example 8 to see how to sketch this function.

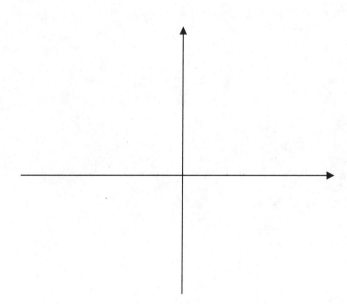

Write down the **Characteristics of Logarithmic Functions** as seen on page 4.3-18.

Section 4.3 Objective 6 Sketching the Graphs of Logarithmic Functions Using Transformations

Do you remember the transformation techniques (how to shift the graphs of functions) that were discussed in Section 2.4? You may want to review these techniques before working on the next set of exercises.

Work through the video that accompanies Example 9:
    Sketch the graph of $f(x) = -\ln(x+2) - 1$

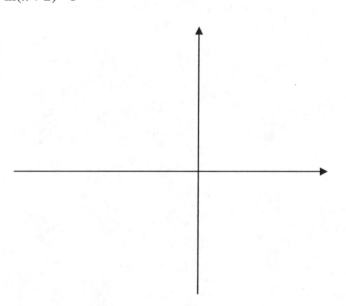

Section 4.3 Objective 7 Finding the Domain of Logarithmic Functions

Look at your graph of $f(x) = -\ln(x+2) - 1$ sketched above. What is the domain of this function? Do you see that the domain is $(-2, \infty)$? To find the domain of a logarithmic function of the form $f(x) = \log_b[g(x)]$, we must find the solution to the inequality $g(x) > 0$.

Example: Find the domain of $f(x) = \log_7 [2x - 5]$. To find the domain, we must solve the inequality $2x - 5 > 0$. Solving this inequality, we get $x > \dfrac{5}{2}$. So the domain of $f(x) = \log_7 [2x - 5]$ is $\left(\tfrac{5}{2}, \infty\right)$.

Now work through the interactive video that accompanies Example 10:

Find the domain of $f(x) = \log_5 \left(\dfrac{2x - 1}{x + 3}\right)$.

Section 4.4 Guided Notebook

Section 4.4 Properties of Logarithms
Work through Objective 1
Work through Objective 2
Work through Objective 3
Work through Objective 4

Section 4.4 Properties of Logarithms

<u>Section 4.4 Objective 1 Using the Product Rule, Quotient Rule, and Power Rule for Logarithms</u>

In Section 4.3, you learned the following properties of logarithms:

General Properties of Logarithms
For $b > 0$ and $b \neq 1$,

1. $\log_b b = 1$

2. $\log_b 1 = 0$

Cancellation Properties of Exponentials and Logarithms
For $b > 0$ and $b \neq 1$,

1. $b^{\log_b b} = b$
2. $\log_b b^x = x$

It is now time to learn some more properties of logarithms. Write down the 3 properties of logarithms seen on page 4.4-2. (You should watch the video proof of each property to get an understanding as to why these properties are true.)

Properties of Logarithms
If $b > 0$ and $b \neq 1$, u and v represent positive real numbers ad r is any real number, then

1.

2.

3.

Section 4.4

Work through the video that accompanies Example 1:

Use the product rule for logarithms to expand each expression. Assume $x > 0$.

a) $\ln(5x)$ b) $\log_2(8x)$

True or False: $\log_b(u+b) = \log_b u + \log_b v$

Try working a "You Try It" on the bottom of page 4.4-3 to see if you understand how to use the product rule for logarithms.

Work through the video that accompanies Example 2:

Use the quotient rule for logarithms to expand each expression. Assume $x > 0$.

a) $\log_5\left(\dfrac{12}{x}\right)$ b) $\ln\left(\dfrac{x}{e^5}\right)$

True or False: $\log_b(u-b) = \log_b u - \log_b v$

True or False: $\log_b u - \log_b v = \dfrac{\log_b u}{\log_b v}$

Try working a "You Try It" on the bottom of page 4.4-4 to see if you understand how to use the quotient rule for logarithms.

Work through the video that accompanies Example 3:

Use the power rule for logarithms to rewrite each expression. Assume $x > 0$.

a) $\log 6^3$ b) $\log_{1/2} \sqrt[4]{x}$

True or False: $\left(\log_b u\right)^r = r \log_b u$

Try working a "You Try It" on the bottom of page 4.4-5 to see if you understand how to use the power rule for logarithms.

Section 4.4 Objective 2 Expanding and Condensing Logarithmic Expressions
Work through the interactive video that accompanies Example 4:

Use properties of logarithms to expand each logarithmic expression as much as possible.

a) $\log_7\left(49x^3\sqrt[5]{y^2}\right)$

b) $\ln\left(\dfrac{\left(x^2-4\right)}{9e^{x^3}}\right)$

Work through the interactive video that accompanies Example 5:

Use properties of logarithms to rewrite each expression as a single logarithm.

a) $\dfrac{1}{2}\log(x-1)-3\log z+\log 5$

b) $\dfrac{1}{3}\left(\log_3 x-2\log_3 y\right)+\log_3 10$

Section 4.4 Objective 3 Solving Logarithmic Equations Using the Logarithm Property of Equality

Why are logarithmic functions one-to-one?

Write down the **Logarithm Property of Equality.**

Work through the interactive video that accompanies Example 6:

Solve the following equations:

a) $\log_7(x-1) = \log_7 12$

b) $2\ln x = \ln 16$

Explain why $x = -4$ is **not** a solution to $2\ln x = \ln 16$.

Section 4.4

<u>Section 4.4 Objective 4 Using the Change of Base Formula</u>
Look at your scientific calculator and locate the $\boxed{\log}$ key and the $\boxed{\ln}$ key. If you are given an expression such as $\log 50$ or $\ln 319$, you can use your calculator to evaluate these expressions. See if you can use your calculator to evaluate these two logarithmic expressions. You should get $\log 50 \approx 1.69897$ and $\ln 319 \approx 5.76519$. How would you use your calculator to evaluate the expression $\log_3 10$? The answer is that you need to change the base from base 3 to base 10 or base e.

Write down the **Change of Base Formula**

You have the skills necessary to prove the Change of Base Formula! Watch the video proof of the Change of Base Formula and write your notes here.

At the end of the video proof, see how to evaluate the expression $\log_3 10$.

$\log_3 10 =$

Work through Example 7:

Approximate the following expressions. Round each to four decimal places.

a) $\log_9 200$ b) $\log_{\sqrt{3}} \pi$

Work through the video that accompanies Example 8:

Use the change of base formula and the properties of logarithms to rewrite as a single logarithm involving base 2.

$$\log_4 x + 3\log_2 y$$

Work through the video that accompanies Example 9:

Use the change of base formula and the properties of logarithms to solve the following equation:

$$2\log_3 x = \log_9 16.$$

Section 4.5 Guided Notebook

Section 4.5 Exponential and Logarithmic Equations

Work through Section 4.5 TTK #1

Work through Section 4.5 TTK #2

Work through Section 4.5 TTK #3

Work through Section 4.5 TTK #5

Work through Section 4.5 TTK #6

Work through Objective 1

Work through Objective 2

GET TO KNOW YOUR CALCULATOR!

In this section, it is crucial that you know how to correctly use your scientific calculator. Test yourself by trying the following calculator exercises. If you are having trouble getting the answers below, please seek help.

1. $e^{-0.0035} \approx .9965$

2. $\dfrac{5e^2 - 4}{e^2 - 1} \approx 5.1565$

3. $\ln 5 \approx 1.6094$

4. $\ln\left(\dfrac{7}{2}\right) \approx 1.2528$

5. $\dfrac{\ln 3 + \ln 5}{\ln 5} \approx 1.6826$

6. $\dfrac{\ln \pi - 3\ln 4}{\ln \pi - 2\ln 4} \approx 1.8516$

7. $\dfrac{\ln\left(\dfrac{17}{3}\right)}{.00235} \approx 738.1281$

8. $\dfrac{\ln\left(\dfrac{77}{131}\right)}{\ln\left(\dfrac{120}{131}\right)} \approx 6.0588$

9. $\dfrac{\ln 2}{12\ln\left(1+\dfrac{.06}{12}\right)} \approx 11.5813$

10. $\dfrac{-\ln 65}{\left(\dfrac{\ln\left(\dfrac{37}{117}\right)}{20}\right)} \approx 72.5188$

Section 4.5 Exponential and Logarithmic Equations

4.5 Things To Know

1. Solving Exponential Equations by Relating the Bases
    Can you solve these two exponential equations?:

a) $8 = \dfrac{1}{16^x}$ b) $\dfrac{1}{27^x} = \left(\sqrt[4]{3}\right)^{x-2}$

Work through the **animation** to refresh your memory.

2. Changing from Exponential to Logarithmic Form

Do you remember how to rewrite an exponential equation as a logarithmic equation? Try rewriting the following exponential equations as equations involving a logarithm. Watch the **video** to see if you are correct.

a) $2^3 = 8$

b) $5^{-2} = \dfrac{1}{25}$

c) $1.1^M = z$

3. Changing from Logarithmic to Exponential Form

Do you remember how to rewrite a logarithmic equation as an exponential equation? Try rewriting the following logarithmic equations as equations involving an exponent. Watch the **video** to see if you are correct.

a) $\log_3 81 = 4$

b) $\log_4 16 = y$

c) $\log_{3/5} x = 2$

5. Expanding and Condensing Logarithmic Expressions
   Expand these logarithmic expressions using properties of logarithms:
   Work through the **interactive video** to see if you are correct.

a) $\log_7\left(49x^3\sqrt[5]{y^2}\right)$

b) $\ln\left(\dfrac{\left(x^2-4\right)}{9e^{x^3}}\right)$

6. Solving Logarithmic Equations Using the Logarithm Property of Equality
   Write down the **Logarithm Property of Equality**:

   Solve these two equations using this property:
   Work through the **interactive video** to see if you are correct.

a) $\log_7(x-1)=\log_7 12$

b) $2\ln x = \ln 16$

<u>Section 4.5 Objective 1 Solving Exponential Equations</u>

Read eText pages 4.5-2 through 4.5-4.

Work through the video that accompanies Example 1: Solve $2^{x+1} = 3$

Write down the summary **Solving Exponential Equations** from page 4.5-5 here:

Work through the interactive video that accompanies Example 2:
    Solve each equation. For part b, round to four decimal places.

a) $3^{x-1} = \left(\dfrac{1}{27}\right)^{2x+1}$

b) $7^{x+3} = 4^{2-x}$

Work through the interactive video that accompanies Example 3:

Solve each equation. Round to four decimal places.

a) $25e^{x-5} = 17$

b) $e^{2x-1} \cdot e^{x+4} = 11$

Section 4.5 Objective 2 Solving Logarithmic Equations

Write down the **logarithm property of equality:**

Write down the three **Properties of Logarithms** as seen on page 4.5-9:

Work through the video that accompanies Example 4:

Solve $2\log_5(x-1) = \log_5 64$

Does the equation above have any extraneous solutions?

What is an extraneous solution? (Click on the definition of an extraneous solution seen on page 4.5-11.)

Write down the steps for **Solving Logarithmic Equations** from page 4.5-11.

Section 4.5

Work through the video that accompanies Example 5 and take notes here. (Refer to the 5 steps to solving logarithmic equations that you wrote on the previous page.)

Solve $\log_4(2x-1)=2$

Work through the interactive video that accompanies Example 6 and take notes here. (Refer to the 5 steps to solving logarithmic equations that you wrote on the previous page.)

Solve $\log_2(x+10)+\log_2(x+6)=5$

Work through Example 7 and take notes here.

Solve $\ln(x-4)-\ln(x-5)=2$. Round to four decimal places.

Section 4.6 Guided Notebook

Section 4.6 Applications of Exponential and Logarithmic Functions

Work through Objective 1
Work through Objective 2
Work through Objective 3
Work through Objective 4

Section 4.6 Applications of Exponential and Logarithmic Equations

Section 4.6 Objective 1 Solving Compound Interest Applications
Complete the two compound interest formulas seen here:

Compound Interest Formulas

Periodic Compound Interest Formula

$$A = \underline{\hspace{4cm}}$$

Continuous Compound Interest Formula

$$A = \underline{\hspace{4cm}}$$

where

A = Total amount after t years
P = Principal (original investment)
r = Interest rate per year
n = Number of times interest is compounded per year
t = Number of years

Work through the video that accompanies Example 1 and take notes here:

How long will it take (in years and months) for an investment to double if it earns 7.5% compounded monthly?

Section 4.6

Work through the video that accompanies Example 2 and take notes here:

> Suppose an investment of $5,000 compounded continuously grew to an amount of
> $5,130.50 in 6 months. Find the interest rate, and then determine how long it will
> take for the investment to grow to $6,000. Round the interest rate to the nearest
> hundredth of a percent and the time to the nearest hundredth of a year.

Section 4.6 Objective 2 Exponential Growth and Decay

Exponential Growth: When a population grows at a rate proportional to the size of the
current population, the following exponential growth function is used. $P(t) = P_0 e^{kt}$ where
$k > 0$ and P_0 (sometimes called "P-not") is the initial population. The graph of this
exponential growth function is seen below.

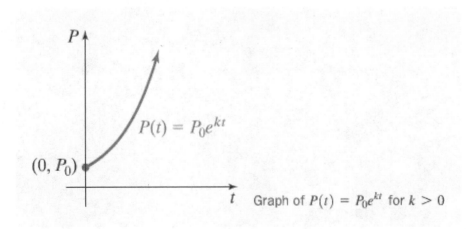

Work through the video that accompanies Example 3:

> The population of a small town grows at a rate proportional to its current size. In 1900, the population was 900. In 1920, the population had grown to 1,600. What was the population of this town in 1950? Round to the nearest whole number.

Exponential Decay

A model that describes the exponential decay of a population, quantity, or amount A, after a certain time, t, is

$$A(t) = A_0 e^{kt}$$

where $A_0 = A(0)$ is the initial quantity and $k < 0$ is a constant called the **relative decay constant**. (*Note*: k is sometimes given as a percent.)

A graph of this exponential decay function is seen on the next page:

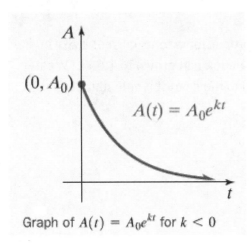

Graph of $A(t) = A_0 e^{kt}$ for $k < 0$

One very useful application of the exponential decay model is **half-life.**

Work through the animation seen on page 4.6-10.

      Define **half-life:**

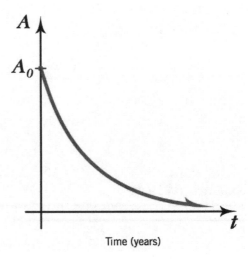

Time (years)

      What is the half-life of Cesium-137?

On the graph on the previous page, insert the relevant time values and corresponding *A* values as seen in this animation. How much Cesium-137 would be left after 120 years? 150 years?

Work through the video that accompanies Example 4:

Suppose that a meteorite is found containing 4% of its original krypton-99. If the half-life of krypton-99 is 80 years, how old is the meteorite? Round to the nearest year.

Section 4.6 Objective 3 Solving Logistic Growth Applications

Logistic Growth:

Another population growth model is used when there are outside factors that affect the population growth. The outside factors may include predators, physical space, disease, etc. This model is called a logistic growth model.

Logistic Growth

A model that describes the logistic growth of a population P at any time t is given by the function

$$P(t) = \frac{C}{1 + Be^{kt}}$$

where B, C, and k are constants with $C > 0$ and $k < 0$.

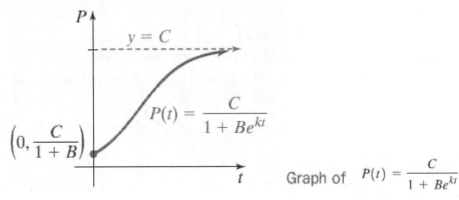

Graph of $P(t) = \dfrac{C}{1 + Be^{kt}}$

In the logistic growth model, what does the number C represent?

Work through the interactive video that accompanies Example 5:

> Ten goldfish were introduced into a small pond. Because of limited food, space, and oxygen, the carrying capacity of the pond is 400 goldfish. The goldfish population at any time t, in days, is modeled by the logistic growth function $F(t) = \dfrac{C}{1 + Be^{kt}}$. If 30 goldfish are in the pond after 20 days,
>
> a) Find B

b) Find k

c) When will the pond contain 250 goldfish? Round to the nearest whole number.

Section 4.6 Objective 4 Using Newton's Law of Cooling
Work through the **animation** that describes Newton's law of cooling.

State **Newton's law of cooling:**

In the Newton's law of cooling model, what does T_0 represent?

In the Newton's law of cooling model, what does S represent?

What is the function that describes Newton's law of cooling?

In the Newton's law of cooling function, what does k represent?

Use the grid below to create a graph similar to the graph in the animation that represents Newton's law of cooling. Be sure to label all relevant information on your graph.

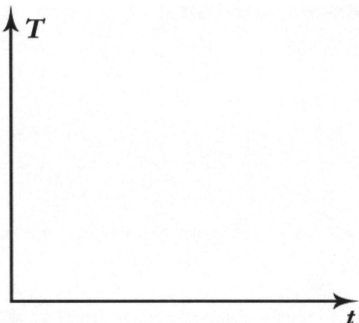

Work through the video that accompanies Example 6:

Suppose that the temperature of a cup of hot tea obeys Newton's law of cooling. If the tea has a temperature of $200°\,F$ when it is initially poured and 1 minute later has cooled to $189°\,F$ in a room that maintains a constant temperature of $69°\,F$, determine when the tea reaches a temperature of $146°\,F$. Round to the nearest minute.

Section 5.1 Guided Notebook

5.1   An Introduction to Angles: Degree and Radian Measure
   Work through Section 5.1 TTK #1
   Work through Section 5.1 Objective 1
   Work through Section 5.1 Objective 2
   Work through Section 5.1 Objective 3
   Work through Section 5.1 Objective 4
   Work through Section 5.1 Objective 5

Section 5.1 An Introduction to Angles: Degree and Radian Measure

<u>**5.1 Things To Know**</u>

1. Sketching the Graph of a Circle (Section 1.2)

Can you sketch the graph of a circle? Try working through a "You Try It" problem or refer to Section 1.2 or watch the video.

<u>Section 5.1 Introduction</u>

What is the definition of a **vertex**?

What is the definition of the **initial side**?

What is the definition of the **terminal side**?

Sketch an angle with positive measure, labeling the vertex, initial side, and terminal side. Do the same for an angle with negative measure.

Section 5.1

What does it mean for an angle to be in **standard position**?

Sketch a coordinate plane and label the **four quadrants**.

Section 5.1 Objective 1 Understanding Degree Measure

In the **degree measure** system, what is the symbol used to indicate a degree? How many degrees are in a one complete counterclockwise rotation?

Sketch three coordinate planes, illustrating angles of 360, 90, and -45 degrees respectively. (See Figures 3, 4, and 5.)

What is the definition of an **acute angle**?

What is the definition of an **obtuse angle**?

What is the definition of a **quadrantal angle**?

What is the term for an angle of exactly 90 degrees?

What is the term for an angle of exactly 180 degrees?

226

What does it mean for angles to be **coterminal**?

Sketch the two coordinate planes illustrating common positive and negative angles as seen in Figure 6.

Work through the video accompanying Example 1 showing all work below.
  Draw each angle in standard position and state the quadrant in which the terminal side of the angle lies or the axis on which the terminal side of the angle lies.
  a. $\theta = 60°$ b. $\alpha = -270°$ c. $\beta = 420°$

Section 5.1

Section 5.1 Objective 2 Finding Coterminal Angles Using Degree Measure

Angles in standard position having the same terminal side are called what?

What is the definition of **Coterminal Angles**?

Starting with a given angle, how can you obtain coterminal angles? (See the definition box on page 5.1-9.)

What notation is used to denote the angle of least positive measure that is coterminal with θ?

Work through the video with Example 2 and show all work below.

Find the angle of least positive measure, θ_C, that is coterminal with $\theta = -697°$.

Section 5.1 Objective 3 Understanding Radian Measure

What is the definition of a **central angle**?

What is the definition of an **intercepted arc**? What variable is typically used to represent it?

What is the definition of a **radian**?

What is the **Relationship between Degrees and Radians**?

Sketch three coordinate planes, illustrating angles of 2π, $\dfrac{\pi}{2}$, and $\dfrac{-\pi}{4}$ radians respectively. (See Figures 10, 11, and 12.)

Sketch two coordinate planes illustrating common positive and negative angles in radians as seen in Figure 13.

Section 5.1

Work through the interactive video accompanying Example 3 showing all work below.
   Draw each angle in standard position and state the quadrant in which the terminal side
   of the angle lies or the axis on which the terminal side of the angle lies.

a. $\theta = \dfrac{\pi}{3}$ b. $\alpha = -\dfrac{3\pi}{2}$ c. $\beta = \dfrac{7\pi}{3}$

Section 5.1 Objective 4 Converting between Degree Measure and Radian Measure

To convert **degrees to radians**, multiply by _____.

To convert **radians to degrees**, multiply by _____.

Work through the interactive video with Example 4 and show all work below.
   Convert each angle given in degree measure into radians.

a.   45°   b. −150°   c. 56°

Work through the interactive video with Example 5 and show all work below.

Convert each angle given in radian measure into degrees. Round to two decimal places if needed.

a. $\dfrac{2\pi}{3}$ radians   b. $-\dfrac{11\pi}{6}$ radians   c. 3 radians

Section 5.1 Objective 5 Finding Coterminal Angles Using Radian Measure

For any angle θ and for any nonzero integer k, we can find a coterminal angle using what expression?

Work through Example 6 and show all work below.

Find three angles that are coterminal with $\theta = \dfrac{\pi}{3}$ using $k = 1$, $k = -1$, and $k = 2$.

Work through the video with Example 7 and show all work below.

Find the angle of least positive measure, θ_C, that is coterminal with $\theta = -\dfrac{21\pi}{4}$.

Section 5.2 Guided Notebook

5.2   Applications of Radian Measure
   Work through Section 5.2 TTK #1
   Work through Section 5.2 Objective 1
   Work through Section 5.2 Objective 2
   Work through Section 5.2 Objective 3

Section 5.2 Applications of Radian Measure

5.2 Things To Know

1. Converting between Degree Measure and Radian Measure (Section 5.1)

Try working through a "You Try It" problem or refer to Section 5.1 or watch the animations.

Section 5.2 Objective 1 Determining the Area of a Sector of a Circle

What is the definition of a **sector** of a circle?

What is the formula for the **Area of a Sector of a Circle**?

Work through Example 1 showing all work below.
   Find the area of the sector of a circle of radius 15 inches formed by a central angle of
   $\theta = \dfrac{\pi}{10}$ radians. Round the answer to two decimal places.

Section 5.2

Use the Caution note at the bottom of page 5.2-4 to fill in the following blanks:

The formula used above is only valid if the angle is given in _____.

An angle given in _____ must first be converted to _____.

Work through the video with Example 2 showing all work below.
   Find the area of the sector of a circle of diameter 21 meters formed by a central angle
   of 135°. Round the answer to two decimal places.

Section 5.2 Objective 2 Computing the Arc Length of a Sector of a Circle

What is the relationship between central angle on a circle of radius r and the length of the
intercepted arc, s? (Hint: See the **Arc Length of a Sector of a Circle** text box on page 5.2-
7.)

Work through Example 3 and show all work below.
   Find the length of the arc intercepted by a central angle of $\theta \dfrac{\pi}{6}$ in a circle of radius
   $r = 22$ centimeters. Round the answer to two decimal places.

Work through the video with Example 4 and show all work below.

A 120° central angle intercepts an arc of 23.4 inches. Calculate the radius of the circle. Round the answer to two decimal places.

Work through the video with Example 5 and show all work below.

Two gears are connected so that the smaller gear turns the larger gear. When the smaller gear with a radius of 3.6 centimeters rotates 300°, how many degrees will the larger gear with a radius of 7.5 cm rotate?

Section 5.2 Objective 3 Understanding Angular Velocity and Linear Velocity

What is the definition of **angular velocity**?

What is the definition of **linear velocity**?

235

Section 5.2

Work through Example 6 showing all work below.

Determine the angular velocity and the linear velocity of a point on the equator of the Earth. Assume that the radius of the Earth at the equator is 3963 miles.

Work through Example 7 showing all work below.

The propeller of a small airplane is rotating 1500 revolutions per minute. Find the angular velocity of the propeller in radians per minute.

Work through Example 8 showing all work below.

In 2008, the old Ferris wheel at the world-famous Santa Monica pier in Santa Monica, California, was replaced with a new solar-powered Ferris wheel that contains thousands of energy-efficient LED (light-emitting diode) lights that can illuminate into many different colors and designs. The wheel is approximately 85 feet in diameter and rotates 2.5 revolutions per minute. Find the linear velocity (in mph) of this new Ferris wheel at the outer edge. Round to two decimal places.

236

What is the **Relationship between Linear Velocity and Angular Velocity**? (Hint: See the text box on page 5.2-16.)

Work through Example 9 showing all work below.

Suppose that the propeller blade from Example 7 is 2.5 meters in diameter. Find the linear velocity in meters per minute for a point located on the tip of the propeller.

Section 5.3 Guided Notebook

5.3   The Unit Circle

Work through Section 5.3 TTK #1-2
Work through Section 5.3 Objective 1
Work through Section 5.3 Objective 2
Work through Section 5.3 Objective 3
Work through Section 5.3 Objective 4
Work through Section 5.3 Objective 5
Work through Section 5.3 Objective 6
Work through Section 5.3 Objective 7

Section 5.3 The Unit Circle

5.3 Things To Know

1. Writing the Standard Form of the Equation of a Circle (Section 1.2)
Try working through a "You Try It" problem or refer to Section 1.2 or watch the video or animation.

2. Converting between Degree Measure and Radian Measure (Section 5.1)
Try working through a "You Try It" problem or refer to Section 5.1 or watch the video.

<u>Section 5.3 Objective 1 Understanding the Definition of the Unit Circle</u>

What is the standard form of the equation of a circle with center (h,k) and radius r ?

What is **The Unit Circle**? (Define it in words and sketch it.)

Fill in the blank: A point (a,b) lies on the graph of the unit circle if and
        only if _____,

Work through the interactive video with Example 1 showing all work below.
        Determine the missing coordinate of a point that lies on the graph of the unit circle
        given the quadrant in which the point is located.

a. $(-\frac{1}{8}, y)$; Quadrant III

b. $(x, -\frac{\sqrt{3}}{2})$; Quadrant IV

c. $(-\frac{1}{\sqrt{2}}, y)$; Quadrant II

239

Section 5.3

<u>Section 5.3 Objective 2 Using Symmetry to Determine Points That Lie on the Unit Circle</u>

What are the three ways in which the unit circle is symmetric?

Work through the video with Example 2 and show all work below.

Verify that the point $(-\frac{1}{8}, -\frac{3\sqrt{7}}{8})$ lies on the graph of the unit circle. Then use symmetry to find three other points that also lie on the graph of the circle.

<u>Section 5.3 Objective 3 Understanding the Unit Circle Definitions of the Trigonometric Functions</u>

The _____ of a sector of the **unit circle** is exactly equal

to the measure of the _____.

What are **The Unit Circle Definitions of the Trigonometric Functions**? (Include the six equations and sketch the graph.)

Section 5.3 Objective 4 Understanding the Fundamental Trigonometric Identities

What are **The Quotient Identities**?

What are **The Reciprocal Identities**?

Work through Example 3 showing all work below.

Given that $\sin t = \dfrac{\sqrt{21}}{5}$ and $\cos t = \dfrac{2}{5}$, find the values of the remaining four trigonometric functions using identities.

What are **The Pythagorean Identities**?

Section 5.3

Work through the video with Example 4 showing all work below.
Use identities to find the exact value of each trigonometric expression.

a. $\tan\dfrac{3\pi}{11} - \dfrac{\sin\dfrac{3\pi}{11}}{\cos\dfrac{3\pi}{11}}$

b. $\dfrac{1}{\cos^2\dfrac{\pi}{9}} - \dfrac{1}{\cot^2\dfrac{\pi}{9}}$

Section 5.3 Objective 5 Using the Unit Circle to Evaluate Trigonometric Functions at Increments of $\pi / 2$

Carefully read pages 5.3-20 through 5.3-26 then fill out Table 1 below:

Table 1

t	$\sin t$	$\cos t$	$\tan t$	$\csc t$	$\sec t$	$\cot t$
$0, 2\pi$						
$\dfrac{\pi}{2}$						
π						
$\dfrac{3\pi}{2}$						

Work through the interactive video with Example 5 and show all work below.

Use the unit circle to determine the value of each expression or state that it is undefined.

a. $\cos 11\pi$

b. $\tan \dfrac{5\pi}{2}$

c. $\csc\left(-\dfrac{3\pi}{2}\right)$

Section 5.3

<u>Section 5.3 Objective 6 Using the Unit Circle to Evaluate Trigonometric Functions for Increments of</u> $\dfrac{\pi}{3}$, $\dfrac{\pi}{6}$, <u>and</u> $\dfrac{\pi}{4}$

Sketch and label **The Unit Circle with Special Points and the Corresponding Special Values of** *t* as seen on page 5.3-30.

Work through the animation with Example 6 and show all work below.
Use the unit circle to determine the following values.

a. $\tan(\frac{7\pi}{3})$ b. $\sin(-\frac{3\pi}{4})$ c. $\sec(\frac{8\pi}{3})$ d. $\csc(-\frac{13\pi}{3})$

Section 5.3 Objective 7 Evaluating Trigonometric Functions Using a Calculator

Work through the video with Example 7 and show all work below.
Evaluate each trigonometric expression using a calculator. Round each answer to four decimal places.

a. $\sin\frac{8\pi}{7}$

b. $\sec\frac{\pi}{5}$

Section 5.4 Guided Notebook

5.4 Right Triangle Trigonometry
Work through Section 5.4 TTK #1-4
Work through Section 5.4 Objective 1
Work through Section 5.4 Objective 2
Work through Section 5.4 Objective 3
Work through Section 5.4 Objective 4
Work through Section 5.4 Objective 5
Work through Section 5.4 Objective 6

Section 5.4 Right Triangle Trigonometry

5.4 Things To Know

1. Understanding Radian Measure (Section 5.1)

Try working through a "You Try It" problem or refer to Section 5.1 or watch the interactive video.

2. Converting between Degree Measure and Radian Measure (Section 5.1)

Try working through a "You Try It" problem or refer to Section 5.1 or watch the animation.

3. Finding Coterminal Angles Using Radian Measure (Section 5.1)

Try working through a "You Try It" problem or refer to Section 5.

4. Understanding the Special Right Triangles (Appendix C.2)

Try working through a "You Try It" problem or refer to Appendix C.2 or watch the animation

Section 5.4 Objective 1 Understanding the Right Triangle Definitions of the Trigonometric Functions

Sketch and label the two triangles seen in Figure 28 on page 5.4-5.

What are **The Right Triangle Definitions of the Trigonometric Functions**?

Work through the interactive video with Example 1 showing all work below.
 Given the right triangle (as seen on page 5.4-8 of the eText), evaluate the six
 trigonometric functions of the acute angle θ.

Section 5.4

Work through the video with Example 2 showing all work below.

If θ is an acute angle of a right triangle and if $\sin \theta = \dfrac{3}{4}$, then find the values of the remaining five trigonometric functions for angle θ.

Section 5.4 Objective 2 Using the Special Right Triangles

Watch the animation that can be found on page 5.4-11, then draw the two special right triangles as seen in Figure 30.

Copy down the trigonometric functions for acute angles $\dfrac{\pi}{6}, \dfrac{\pi}{4}, \dfrac{\pi}{3}$ as seen in Table 2.

Table 2

θ	$\dfrac{\pi}{6}(30°)$	$\dfrac{\pi}{4}(45°)$	$\dfrac{\pi}{3}(60°)$
$\sin \theta$			
$\cos \theta$			
$\tan \theta$			

Work through the video with Example 3 and show all work below.

Determine the value of $\csc \dfrac{\pi}{6} + \cot \dfrac{\pi}{4}$.

Work through Example 4 and show all work below.

Determine the measure of the acute angle θ for which $\sec \theta = 2$.

<u>Section 5.4 Objective 3 Understanding Cofunctions</u>

What does it mean for two angles to be **complementary**?

What are the **Cofunction Identities**?

Section 5.4

Work through the interactive video with Example 5 and show all work below.

a. Rewrite the expression $(\cot(\frac{\pi}{2}-\theta))\cos\theta$ as one of the six trigonometric functions of acute angle θ.

b. Determine the exact value of $\sec 55° \csc 35° - \tan 55° \cot 35°$.

Section 5.4 Objective 4 Understanding the Four Families of Special Angles

What is **The Quadrantal Family of Angles**? Sketch the angles shown in Figure 32.

What is **The $\frac{\pi}{3}$ Family of Angles**? Sketch the angles shown in Figure 33.

What is **The $\frac{\pi}{6}$ Family of Angles**? Sketch the angles shown in Figure 34.

What is **The $\frac{\pi}{4}$ Family of Angles**? Sketch the angles shown in Figure 35.

Work through the interactive video with Example 6 showing all work below.

Each of the given angles belongs to one of the four families of special angles. Determine the family of angles for which it belongs, sketch the angle, and then determine the angle of least positive measure, θ_C, coterminal with the given angle.

a. $\theta = \dfrac{29\pi}{6}$ b. $\theta = \dfrac{14\pi}{2}$ c. $\theta = -\dfrac{18\pi}{4}$ d. $\theta = \dfrac{11\pi}{4}$

e. $\theta = \dfrac{14\pi}{6}$ f. $\theta = 420°$ g. $\theta = -495°$

Section 5.4 Objective 5 Understanding the Definitions of the Trigonometric Functions of General Angles

Work through the animation as seen on page 5.4-28 and take notes here.

What are **The General Angle Definitions of the Trigonometric Functions**?

Under what conditions will the following trigonometric functions be undefined (if ever)?

$$\tan \theta = \frac{y}{x} \text{ and } \sec \theta = \frac{r}{x}:$$

$$\csc \theta = \frac{r}{y} \text{ and } \cot \theta = \frac{x}{y}:$$

Section 5.4

Work through the video with Example 7 and show all work below.

     Suppose that the point $(-4, -6)$ is on the terminal side of an angle θ. Find the six trigonometric functions of θ.

<u>Section 5.4 Objective 6 Finding the Values of the Trigonometric Functions of Quadrantal Angles</u>

Watch the video as seen on page 5.4-33, then fill in Table 3.

Table 3

θ	$\sin \theta$	$\cos \theta$	$\tan \theta$	$\csc \theta$	$\sec \theta$	$\cot \theta$
0						
$\dfrac{\pi}{2}$						
π						
$\dfrac{3\pi}{2}$						

254

Work through the interactive video accompanying Example 8 showing all work below.
Without using a calculator, determine the value of the trigonometric function or state that the value is undefined.

a. $\cos(-11\pi)$ b. $\csc(-270°)$ c. $\tan(\dfrac{13\pi}{2})$

d. $\sin(540°)$ e. $\cot(-\dfrac{7\pi}{2})$

Section 5.4 Objective 7 Understanding the Signs of the Trigonometric Functions

The sign of each trigonometric function is determined by the _____ in which the terminal side of the angle lies.

Which trigonometric functions are positive for all angles with a terminal side lying in the following quadrants?

Quadrant I:

Quadrant II:

Quadrant III:

Quadrant IV:

What acronym can help us remember the signs of the trigonometric functions for angles whose terminal side lies in one of the four quadrants?

Sketch the diagram shown in Figure 46.

Work through the video with Example 9 and show all work below.

Suppose θ is a positive angle in standard position such that $\sin\theta < 0$ and $\sec\theta > 0$.

a. Determine the quadrant in which the terminal side of angle θ lies.

b. Find the value of $\tan\theta$ if $\sec\theta = \sqrt{5}$.

Section 5.4 Objective 8 Determining Reference Angles

What is the definition of the **Reference Angle**?

The measure of the _____ θ_R depends on the quadrant in

which the _____ of θ_C lies.

Sketch the four cases as seen on pages 5.4-46 and 5.4-47 of the eText.

Work through the interactive video with Example 10 and show all work below.
For each of the given angles, determine the reference angle.

a. $\theta = \dfrac{5\pi}{3}$ b. $\theta = \dfrac{11\pi}{4}$ c. $\theta = -\dfrac{25\pi}{6}$ d. $\theta = \dfrac{16\pi}{6}$

Work through the interactive video with Example 11 and show all work below.
For each of the given angles, determine the reference angle.

a. $\theta = \dfrac{5\pi}{8}$ b. $\theta = \dfrac{22\pi}{9}$ c. $\theta = -\dfrac{5\pi}{7}$

Section 5.4

Work through the interactive video with Example 12 and show all work below.
For each of the given angles, determine the reference angle.
a. $\theta = 225°$ b. $\theta = -233°$ c. $\theta = 510°$

Section 5.4 Objective 9 Evaluating Trigonometric Functions of Angles Belonging to the pi/3, pi/6, or pi/4 Families

Work through the interactive video with Example 13 and show all work below.
Find the values of the six trigonometric functions for $\theta = \dfrac{7\pi}{4}$.

What are the four **Steps for Evaluating Trigonometric Functions of Angles Belonging to the $\dfrac{\pi}{3}$, $\dfrac{\pi}{6}$, or $\dfrac{\pi}{4}$ Families?**

Step 1:

Step 2:

Step 3:

Step 4:

260

Work through the interactive video with Example 14 and show all work below.

Find the exact value of each trigonometric expression without using a calculator.

a. $\sin(\dfrac{7\pi}{6})$

b. $\cot(-\dfrac{22\pi}{3})$

c. $\tan(\dfrac{11\pi}{4})$

d. $\cos(\frac{11\pi}{3})$

e. $\sec(\frac{5\pi}{6})$

f. $\csc(-\frac{7\pi}{6})$

Section 6.1 Guided Notebook

6.1 The Graphs of Sine and Cosine

Work through Section 6.1 TTK #1-6
Work through Section 6.1 Objective 1
Work through Section 6.1 Objective 2
Work through Section 6.1 Objective 3
Work through Section 6.1 Objective 4
Work through Section 6.1 Objective 5

Section 6.1 The Graphs of Sine and Cosine

6.1 Things To Know

1. Determining Whether a Function is Even, Odd, or Neither (Section 2.2)
Try working through a "You Try It" problem or refer to Section 2.2 or watch the video.

2. Using Vertical Stretches and Compressions to Graph Functions (Section 2.4)
Try working through a "You Try It" problem or refer to Section 2.4 or watch the video.

3. Using Horizontal Stretches and Compressions to Graph Functions (Section 2.4)
Try working through a "You Try It" problem or refer to Section 2.4 or watch the video.

4. Using the Unit Circle to Evaluate Trigonometric Functions at Increments of $\frac{\pi}{2}$ (Section 5.3)
Try working through a "You Try It" problem or refer to Section 5.3 or watch the video.

5. Use the Unit Circle to Evaluate Trigonometric Functions for Increments of $\frac{\pi}{3}$, $\frac{\pi}{6}$, and $\frac{\pi}{4}$ Families (Section 5.3)
Try working through a "You Try It" problem or refer to Section 6.5 or watch the video.

Section 6.1

Section 6.1 Objective 1 Understanding the Graph of the Sine Function and Its Properties

Write down the tables as seen in Tables 1 through 4 and fill in the missing information.

What is the definition of a **Periodic Function**?

Write down the **Characteristics of the Sine Function** (as seen on page 6.1-8)

Work through Example 1 showing all work below.

Using the graph of $y = \sin x$, list all values of x on the interval $[-3\pi, \frac{7\pi}{4}]$ that satisfy the ordered pair $(x, 0)$.

Work through Example 2 showing all work below.

Use the periodic property of $y = \sin x$ to determine which of the following expressions is equivalent to $\sin(\frac{23\pi}{6})$.

i. $\sin(\frac{\pi}{6})$ ii. $\sin(\frac{5\pi}{6})$ iii. $\sin(\frac{11\pi}{6})$ iv. $\sin(\frac{13\pi}{6})$

Work through Example 3 showing all work below.

Use the fact that $y = \sin x$ is an odd function to determine which of the following expressions is equivalent to $-\sin(\frac{9\pi}{16})$.

i. $\sin(-\frac{9\pi}{16})$ ii. $\sin(\frac{9\pi}{16})$ iii. $-\sin(-\frac{9\pi}{16})$

Section 6.1 Objective 2 Understanding the Graph of the Cosine Function and Its Properties

Write down the tables as seen in Tables 5 through 8 and fill in the missing information.

Write down the **Characteristics of the Cosine Function** (as seen on page 6.1-16

Work through Example 4 showing all work below.

Using the graph of $y = \cos x$, list all values of x on the interval $[-\pi, 2\pi]$ that satisfy the ordered pair $(x, \frac{1}{2})$.

Write down **The Five Quarter Points of** $y = \sin x$ **and** $y = \cos x$ and sketch the graphs (as seen on page 6.1-18).

Section 6.1 Objective 3 Sketching Graphs of the From $y=A\sin x$ and $y=A\cos x$

What is the definition of **amplitude**?

Section 6.1

Work through the video with Example 5 showing all work below.

Determine the amplitude and range of $y = -\dfrac{2}{3}\cos x$ and then sketch the graph.

Section 6.1 Objective 4 Sketching Graphs of the Form $y=\sin(Bx)$ and $y=\cos(BX)$

How do you **Determine the Period of** $y = \sin(Bx)$ **and** $y = \cos(Bx)$

Work through the interactive video with Example 6 and show all work below.
Determine the period and sketch the graph of each function.

a. $y = \sin(2x)$ b. $y = \cos(\frac{1}{2}x)$ c. $y = \sin(\pi x)$

Work through the interactive video with Example 7 and show all work below.
Determine the period and sketch the graph of each function.

a. $y = \sin(-2x)$ b. $y = \cos(-\frac{1}{2}x)$ c. $y = \sin(-\pi x)$

269

Section 6.1

Section 6.1 Objective 5 Sketching Graphs of the Form $y=A\sin(Bx)$ and $y=A\cos(BX)$

What are the six **Steps for Sketching Functions of the Form $y=A\sin(Bx)$ and $y=A\cos(BX)$?**

Step 1.

Step 2.

Step 3.

Step 4.

Step 5.

Step 6.

Work through the interactive video with Example 8 and show all work below.
    Use the six-step process outlined in this section to sketch each graph.

    a. $y = 3\sin(4x)$ b. $y = -2\cos(\frac{1}{3}x)$ c. $y = -6\sin(-\frac{\pi x}{2})$

Section 6.2 Guided Notebook

6.2 More on Graphs of Sine and Cosine: Phase Shift
    Work through Section 6.2 TTK #1-5
    Work through Section 6.2 Objective 1
    Work through Section 6.2 Objective 2
    Work through Section 6.2 Objective 3

Section 6.2 More on Graphs of Sine and Cosine: Phase Shift

6.2 Things To Know

1. Using Vertical Shifts to Graph Functions (Section 2.4)
Try working through a "You Try It" problem or refer to Section 2.4 or watch the animation.

2. Using Horizontal Shifts to Graph Functions (Section 2.4)
Try working through a "You Try It" problem or refer to Section 2.4 or watch the animation.

3. Sketching Graphs of the Form $y=A\sin x$ and $y=A\cos x$ (Section 6.1)
Try working through a "You Try It" problem or refer to Section 6.1 or watch the video.

4. Sketching Graphs of the Form $y=\sin(Bx)$ and $y=\cos(Bx)$ (Section 6.1)
Try working through a "You Try It" problem or refer to Section 6.1 or watch the interactive video.

5. Sketching Graphs of the Form $y=A\sin(Bx)$ and $y=A\cos(Bx)$ (Section 6.1)
Try working through a "You Try It" problem or refer to Section 6.1 or watch the interactive video.

Section 6.2

Section 6.2 Objective 1 Sketching Graphs of the Form $y=\sin(x-C)$ and $y=\cos(x-C)$

What is the **phase shift**?

Write down the properties and sketch the graphs of $y=\sin(x-C)$ and $y=\cos(x-C)$ as seen on page 6.2-5.

Work through the interactive video with Example 1 showing all work below.
Determine the phase shift and sketch the graph of each function.

a. $y = \cos(x - \pi)$ b. $y = \sin(x + \dfrac{\pi}{2})$

What is the **Relationship between Graphs of Sine and Cosine Functions** (as seen on page 6.2-11)?

Section 6.2 Objective 2 Sketching Graphs of the Form $y=A\sin(Bx-C)$ and $y=A\cos(Bx-C)$

What are the seven **Steps for Sketching Functions of the Form $y=A$sin(Bx-C) and $y=A$cos(Bx-C)?**

Step 1.

Step 2.

Step 3.

Step 4.

Step 5.

Step 6.

Step 7.

Section 6.2

Work through the interactive video with Example 2 showing all work below.
Sketch the graph of each function.

a. $y = 3\sin(2x - \pi)$

b. $y = -2\cos(3x + \dfrac{\pi}{2})$

c. $y = 3\sin(\pi - 2x)$

d. $y = -2\cos(-3x + \dfrac{\pi}{2})$

Section 6.2 Objective 3 Sketching Graphs of the Form $y=A\sin(Bx-C)+D$ and $y=A\cos(Bx-C)+D$

What are the seven **Steps for Sketching Functions of the Form $y=A\sin(Bx-C)+D$ and $y=A\cos(Bx-C)+D$**?

Step 1.

Step 2.

Step 3.

Step 4.

Step 5.

Step 6.

Step 7.

Section 6.2

Work through the interactive video with Example 3 showing all work below.
    Sketch the graph of each function.

a. $y = 3\sin(2x - \frac{\pi}{2}) - 1$ b. $y = 4 - \cos(-\pi x + 2)$

Section 6.3 Guided Notebook

6.3   The Graphs of the Tangent, Cosecant, Secant, and Cotangent Functions
Work through Section 6.3 TTK #1-4
Work through Section 6.3 Objective 1
Work through Section 6.3 Objective 2
Work through Section 6.3 Objective 3
Work through Section 6.3 Objective 4
Work through Section 6.3 Objective 5
Work through Section 6.3 Objective 6

Section 6.3 The Graphs of the Tangent, Cosecant, Secant, and Cotangent Functions

6.3 Things To Know

1. Using the Unit Circle to Evaluate Trigonometric Functions at Increments of $\dfrac{\pi}{2}$ (Section 5.3)
Try working through a "You Try It" problem or refer to Section 5.3 or watch the video.

2 Using the Unit Circle to Evaluate Trigonometric Functions for Increments of $\dfrac{\pi}{3}$, $\dfrac{\pi}{6}$, and $\dfrac{\pi}{4}$ Families (Section 5.3)
Try working through a "You Try It" problem or refer to Section 6.5 or watch the video.

3. Sketching Graphs of the Form $y=A\sin(Bx)$ and $y=A\cos(Bx)$ (Section 6.1)
Try working through a "You Try It" problem or refer to Section 6.1 or watch the interactive video.

4. Sketching Graphs of the Form $y=A\sin(Bx-C)+D$ and $y=A\cos(Bx-C)+D$ (Section 6.2)
Try working through a "You Try It" problem or refer to Section 6.2 or watch the interactive video.

<u>Section 6.3 Objective 1 Understanding the Graph of the Tangent Function and Its Properties</u>

What is the **principal cycle** of the graph of $y = \tan x$? Sketch its graph.

What are the three special points in each cycle of the graph of $y = \tan x$ that will help us to sketch the graph? Go back to the sketch above and add these special points.

Write down the **Characteristics of the Tangent Function** (as seen on page 6.3-9)

Work through the video with Example 1 showing all work below.

List all the halfway points of $y = \tan x$ on the interval $[-\pi, \frac{5\pi}{2}]$ that have a y-coordinate of -1.

Section 6.3 Objective 2 Sketching Functions of the Form $y=A\tan(Bx-C)+D$

What are the six Steps for Sketching Functions of the Form $y=A\tan(Bx-C)+D$?

Step 1.

Step 2.

Step 3.

Step 4.

Step 5.

Step 6.

Section 6.3

Work through the interactive video with Example 2 showing all work below.

For each function, determine the interval for the principal cycle. Then for the principal cycle, determine the equations of the vertical asymptotes, the coordinates of the center points, and the coordinates of the halfway points. Sketch the graph.

a. $y = \tan(x - \frac{\pi}{6})$ b. $y = 4\tan(\pi - 2x) + 3$ c. $y = \frac{1}{2}\tan(3x) - 1$

Section 6.3 Objective 3 Understanding the Graph of the Cotangent Function and Its Properties

What are the **Characteristics of the Cotangent Function**?

Work through the video with Example 3 showing all work below.

List all points on the graph of $y = \cot x$ on the interval $[-2\pi, 2\pi]$ that have a y-coordinate of $-\sqrt{3}$.

Section 6.3

Section 6.3 Objective 4 Sketching Functions of the Form $y=A\cot(Bx-C)+D$

What are the six **Steps for Sketching Functions of the Form** $y=A\cot(Bx-C)+D$?

Step 1.

Step 2.

Step 3.

Step 4.

Step 5.

Step 6.

Work through the interactive video with Example 4 and show all work below.

For each function, determine the interval for the principal cycle. Then for the principal cycle, determine the equations of the vertical asymptotes, the coordinates of the center points, and the coordinates of the halfway points. Sketch the graph.

a. $y = \cot(2x + \pi) + 1$ b. $y = -3\cot(x - \dfrac{\pi}{4})$

<u>Section 6.3 Objective 5 Understanding the Graphs of the Cosecant and Secant Functions and Their Properties</u>

What are the **Characteristics of the Cosecant Function**?

Section 6.3

What are the **Characteristics of the Secant Function**?

Section 6.3 Objective 6 Sketching Functions of the Form $y=A\csc(Bx-C)+D$ and $y=A\sec(Bx-C)+D$

What are the four **Steps for Sketching Functions of the Form**
$y=A\csc(Bx-C)+D$ **and** $y=A\sec(Bx-C)+D$?

Step 1.

Step 2.

Step 3.

Step 4.

284

Work through the interactive video with Example 5 and show all work below.

Determine the equations of the vertical asymptotes and all relative maximum and relative minimum points of two cycles of each function and then sketch its graph.

a. $y = -2\csc(x+\pi)$

b. $y = -\csc(\pi x) - 2$

c. $y = 3\sec(\pi - x)$

d. $y = \sec(2x + \dfrac{\pi}{2}) + 1$

Section 6.4 Guided Notebook

6.4   Inverse Trigonometric Functions I
Work through Section 6.4 TTK #1-11
Work through Section 6.4 Objective 1
Work through Section 6.4 Objective 2
Work through Section 6.4 Objective 3

Section 6.4 Inverse Trigonometric Functions I

6.4 Things To Know

1. Determining Whether a Function is One-to-One Using the Horizontal Line Test (Section 2.6)
Try working through a "You Try It" problem or refer to Section 2.6 or watch the video.

2. Understanding the Definition of an Inverse Function (Section 2.6)
Try working through a "You Try It" problem or refer to Section 2.6 or watch the video.

3. Sketching the Graph of an Inverse Function (Section 2.6)
Try working through a "You Try It" problem or refer to Section 2.6 or watch the animation.

4. Understanding the Special Right Triangles (Appendix C.2)
Try working through a "You Try It" problem or refer to Appendix C.2 or watch the animation.

5. Understanding the Right Triangle Definitions of the Trigonometric Functions (Section 5.4)
Try working through a "You Try It" problem or refer to Section 6.4 or watch the video.

6. Understanding the Signs of the Trigonometric Functions (Section 5.4)
Try working through a "You Try It" problem or refer to Section 5.4 or watch the video.

7. Determining Reference Angles (Section 5.4)
Try working through a "You Try It" problem or refer to Section 5.4 or watch the interactive video.

8. Evaluating Trigonometric Functions of Angles Belonging to the $\frac{\pi}{3}$, $\frac{\pi}{6}$, or $\frac{\pi}{4}$ Families (Section 5.4)
Try working through a "You Try It" problem or refer to Section 5.4 or watch the interactive video.

9. Understanding the Graph of the Sine Function and Its Properties (Section 6.1)
Try working through a "You Try It" problem or refer to Section 6.1 or watch the video.

10. Understanding the Graph of the Cosine Function and Its Properties (Section 6.1)
Try working through a "You Try It" problem or refer to Section 6.1 or watch the video.

11. Understanding the Graph of the Tangent Function and Its Properties (Section 6.3)
Try working through a "You Try It" problem or refer to Section 7.3.

<u>Section 6.4 Objective 1 Understanding and Finding the Exact and Approximate Values of the Inverse Sine Function</u>

What is the definition of the **Inverse Sine Function**? (Define in words and sketch the graph.)

What are the four **Steps for Determining the Exact Value of** $\sin^{-1} x$ **?**

Step 1.

Step 2.

Step 3.

Step 4.

287

Section 6.4

Work through the interactive video with Example 1 showing all work below.
  Determine the exact value of each expression.

  a. $\sin^{-1}\left(\frac{1}{2}\right)$ b. $\sin^{-1}\left(-\frac{\sqrt{3}}{2}\right)$

Work through Example 2 showing all work below.
  Use a calculator to approximate each value, or state that the value does not exist.
  a. $\sin^{-1}(.7)$ b. $\sin^{-1}(-.95)$ c. $\sin^{-1}(3)$

Section 6.4 Objective 2 Understanding and Finding the Exact and Approximate Values of the Inverse Cosine Function

What is the definition of the **Inverse Cosine Function**? (Define and sketch.)

288

What are the four **Steps for Determining the Exact Value of** $\cos^{-1} x$?

Step 1.

Step 2.

Step 3.

Step 4.

Work through the interactive video with Example 3 showing all work below.
    Determine the exact value of each expression.

    a. $\cos^{-1}(1)$ b. $\cos^{-1}(-\frac{1}{\sqrt{2}})$

Section 6.4

Work through Example 4 showing all work below.
Use a calculator to approximate each value, or state that the value does not exist.
a. $\cos^{-1}(1.5)$ b. $\cos^{-1}(-.25)$

Section 6.4 Objective 3 Understanding and Finding the Exact and Approximate Values of the Inverse Tangent Function

What is the definition the **Inverse Tangent Function**? (Define and sketch.)

What are the four **Steps for Determining the Exact Value of** $\tan^{-1} x$?

Step 1.

Step 2.

Step 3.

Step 4.

Work through the interactive video with Example 5 showing all work below.
Determine the exact value of each expression.

a. $\tan^{-1}(\dfrac{1}{\sqrt{3}})$ b. $\tan^{-1}(-\dfrac{1}{\sqrt{3}})$

Work through Example 6 showing all work below.
Use a calculator to approximate the value of $\tan^{-1}(20)$, or state that the value does not exist.

Section 6.5 Guided Notebook

6.5   Inverse Trigonometric Functions II
Work through Section 6.5 TTK #1-9
Work through Section 6.5 Objective 1
Work through Section 6.5 Objective 2
Work through Section 6.5 Objective 3
Work through Section 6.5 Objective 4

Section 6.5 Inverse Trigonometric Functions II

6.5 Things To Know

1. Understanding the Composition Cancellation Equations (Section 2.6)
Try working through a "You Try It" problem or refer to Section 2.6 or watch the video.

2. Understanding the Special Right Triangles (Appendix C.2)
Try working through a "You Try It" problem or refer to Appendix C.2 or watch the animation.

3. Understanding the Right Triangle Definitions of the Trigonometric Functions (Section 5.4)
Try working through a "You Try It" problem or refer to Section 5.4 or watch the video.

4. Understanding the Signs of the Trigonometric Functions (Section 5.4)
Try working through a "You Try It" problem or refer to Section 5.4 or watch the video.

5. Determining Reference Angles (Section 5.4)
Try working through a "You Try It" problem or refer to Section 5.4 or watch the interactive video.

6. Evaluating Trigonometric Functions of Angles Belonging to the $\frac{\pi}{3}$, $\frac{\pi}{6}$, or $\frac{\pi}{4}$ Families (Section 5.4)
Try working through a "You Try It" problem or refer to Section 5.4 or watch the video.

7. Understanding the Inverse Sine Function (Section 6.4)
Try working through a "You Try It" problem or refer to Section 6.4 or watch the video.

8. Understanding the Inverse Cosine Function (Section 6.4)
Try working through a "You Try It" problem or refer to Section 6.4 or watch the video.

9. Understanding the Inverse Tangent Function (Section 6.4)
Try working through a "You Try It" problem or refer to Section 6.4 or watch the video.

Section 6.5 Objective 1 Evaluating Composite Functions Involving Inverse Trigonometric Functions of the Form $f \circ f^{-1}$ and $f^{-1} \circ f$

What are the **Cancellation Equations for Compositions of Inverse Trigonometric Functions**?

Work through the interactive video with Example 1 showing all work below.
Find the exact value of each expression or state that it does not exist.

a. $\sin(\sin^{-1}\frac{1}{2})$ b. $\cos(\cos^{-1}\frac{3}{2})$ c. $\tan(\tan^{-1}(8.2))$ d. $\sin(\sin^{-1}(1.3))$

Work through the interactive video with Example 2 showing all work below.
Find the exact value of each expression or state that it does not exist.

a. $\sin^{-1}(\sin\frac{\pi}{6})$ b. $\cos^{-1}(\cos\frac{2\pi}{3})$ c. $\sin^{-1}(\sin\frac{4\pi}{3})$ d. $\tan^{-1}(\tan\frac{7\pi}{10})$

Section 6.5

Work through the interactive video with Example 3 showing all work below.
   Find the exact value of each expression or state that it does not exist.

   a. $\cos(\tan^{-1}\sqrt{3})$ b. $\csc(\cos^{-1}(-\frac{\sqrt{3}}{2}))$ c. $\sec(\sin^{-1}(-\frac{\sqrt{5}}{8}))$

Work through the interactive video with Example 4 showing all work below.
   Find the exact value of each expression or state that it does not exist.

   a. $\sin^{-1}(\cos(-\frac{2\pi}{3}))$ b. $\cos^{-1}(\sin\frac{\pi}{7})$

Section 6.5 Objective 3 Understanding the Inverse Cosecant, Inverse Secant, and Inverse Cotangent Functions

What is the definition of the **Inverse Cosecant Function**? (Define and sketch.)

What is the definition of the **Inverse Secant Function**? (Define and sketch.)

What is the definition of the **Inverse Cotangent Function**? (Define and sketch.)

Work through Example 5 showing all work below.

Find the exact value of $\sec^{-1}(-\sqrt{2})$ or state that it does not exist.

Work through Example 6 showing all work below.

Use a calculator to approximate each value or state that the value does not exist.

a. $\sec^{-1}(5)$ b. $\cot^{-1}(-10)$ c. $\csc^{-1}(0.4)$

Section 6.5 Objective 4 Writing Trigonometric Expressions as Algebraic Expressions

Work through the video with Example 7 showing all work below.

Rewrite the trigonometric expression $\sin(\tan^{-1} u)$ as an algebraic expression involving the variable u. Assume that $\tan^{-1} u$ represents an angle whose terminal side is located in Quadrant I.

Section 7.1 Guided Notebook

7.1   Trigonometric Identities
   Work through Section 7.1 TTK #1-3
   Work through Section 7.1 Objective 1
   Work through Section 7.1 Objective 2
   Work through Section 7.1 Objective 3
   Work through Section 7.1 Objective 4
   Work through Section 7.1 Objective 5
   Work through Section 7.1 Objective 6
   Work through Section 7.1 Objective 7
   Work through Section 7.1 Objective 8

Section 7.1 Trigonometric Identities

7.1 Things To Know

1. Understanding the Quotient Identities for Acute Angles (Section 5.3)
Try working through a "You Try It" problem or refer to Section 5.3 or watch the video.

2. Understanding the Reciprocal Identities for Acute Angles (Section 5.3)
Try working through a "You Try It" problem or refer to Section 5.3 or watch the video.

3. Understanding the Pythagorean Identities for Acute Angles (Section 5.3)
Try working through a "You Try It" problem or refer to Section 5.3 or watch the video.

Section 7.1 Objective 1 Reviewing the Fundamental Identities

What are the two **Quotient Identities**?

What are the six **Reciprocal Identities**?

What are the three **Pythagorean Identities**?

What are the four **Odd Properties**?

What are the two **Even Properties**?

Section 7.1 Objective 2 Substituting Known Identities to Verify an Identity

Work through the interactive video with Example 1 showing all work below.
Verify each identity.
a. $\tan x \cot x = 1$
b. $\sec^2 3x + \cot^2 3x - \tan^2 3x = \csc^2 3x$
c. $(5 \sin y + 2 \cos y)^2 + (5 \cos y - 2 \sin y)^2 = 29$

Section 7.1

Section 7.1 Objective 3 Changing to Sines and Cosines to Verify an Identity

Work through the interactive video with Example 2 showing all work below.
   Verify each identity.
   a. $\sin^2 t = \tan t \cot t - \cos^2 t$

   b. $\dfrac{\sec\theta \csc\theta}{\cot\theta} = \sec^2\theta$

   c. $\dfrac{\cos(-\theta)}{\sec\theta} + \sin(-\theta)\csc\theta = -\sin^2\theta$

Section 7.1 Objective 4 Factoring to Verify an Identity

Write down the following special factoring formulas:
Difference of Two Squares:

Perfect Square Formulas:

Sum of Two Cubes:

Difference of Two Cubes:

Work through the interactive video with Example 3 and show all work below.
Verify each trigonometric identity.
a. $\sin x - \cos^2 x \sin x = \sin^3 x$

b. $\dfrac{\tan^3 \alpha - 1}{\tan \alpha - 1} = \sec^2 \alpha + \tan \alpha$

c. $\dfrac{8\sin^2\theta - 2\sin\theta - 3}{1 + 2\sin\theta} = 4\sin\theta - 3$

Section 7.1 Objective 5 Separating a Single Quotient into Multiple Quotients to Verify an Identity

Work through the video with Example 4 and show all work below.

Verify the trigonometric identity $\dfrac{\sin\alpha + \cos\alpha}{\cos\alpha} - \dfrac{\sin\alpha + \cos\alpha}{\sin\alpha} = \tan\alpha - \cot\alpha$.

Section 7.1 Objective 6 Combining Fractional Expressions to Verify an Identity

Work through the video with Example 5 and show all work below.

Verify the trigonometric identity $\dfrac{1-\csc\theta}{\cot\theta} - \dfrac{\cot\theta}{1-\csc\theta} = 2\tan\theta$.

Section 7.1 Objective 7 Multiplying by Conjugates to Verify Identities

Work through the video with Example 6 and show all work below.

Verify the trigonometric identity $\dfrac{\sin\theta}{\csc\theta+1} = \dfrac{1-\sin\theta}{\cot^2\theta}$.

Section 7.1

Section 7.1 Objective 8 Summarizing the Techniques for Verifying Identities

What are the six techniques described in **A Summary for Verifying Trigonometric Identities**?

1.

2.

3.

4.

5.

6.

Work through the interactive video with Example 7 and show all work below.
Verify each trigonometric identity.

a. $\dfrac{\sin^2 t + 6\sin t + 9}{\sin t + 3} = \dfrac{3\csc t + 1}{\csc t}$

b. $\dfrac{2\csc\theta}{\sec\theta} + \dfrac{\cos\theta}{\sin\theta} = 3\cot\theta$

c. $\dfrac{1-\sin\theta}{\cos\theta} + \dfrac{\cos\theta}{1-\sin\theta} = 2\sec\theta$

Section 7.2 Guided Notebook

7.2   The Sum and Difference Formulas
Work through Section 7.2 TTK #1-4
Work through Section 7.2 Objective 1
Work through Section 7.2 Objective 2
Work through Section 7.2 Objective 3
Work through Section 7.2 Objective 4
Work through Section 7.2 Objective 5

Section 7.2 The Sum and Difference Formulas

<u>**7.2 Things To Know**</u>

1. Understanding Cofunctions (Section 5.4)
Try working through a "You Try It" problem or refer to Section 5.4 or watch the video.

2. Evaluating Trigonometric Functions of Angles Belonging to the $\frac{\pi}{3}$, $\frac{\pi}{6}$, or $\frac{\pi}{4}$ Families
(Section 5.4)
Try working through a "You Try It" problem or refer to Section 5.4 or watch the interactive video.

3. Finding the Exact and Approximate Values of an Inverse Sine Expression (Section 6.4)
Try working through a "You Try It" problem or refer to Section 64 or watch the interactive video.

4. Finding the Exact and Approximate Values of an Inverse Cosine Expression (Section 6.4)
Try working through a "You Try It" problem or refer to Section 6.4 or watch the interactive video.

<u>Section 7.2 Objective 1 Understanding the Sum and Difference Formulas for the Cosine Function</u>

What are **The Sum and Difference Formulas for the Cosine Function**?

Work through the interactive video with Example 1 showing all work below.

Find the exact value of each trigonometric expression without the use of a calculator.

a. $\cos(\dfrac{2\pi}{3} + \dfrac{3\pi}{4})$

b. $\cos(225° - 150°)$

Work through the interactive video with Example 2 showing all work below.

Find the exact value of each trigonometric expression without the use of a calculator.

a. $\cos(\dfrac{7\pi}{12})\cos(\dfrac{5\pi}{12}) + \sin(\dfrac{7\pi}{12})\sin(\dfrac{5\pi}{12})$

b. $\cos(\dfrac{7\pi}{12})$

c. $\cos(-75°)$

307

Section 7.2

Work through the video with Example 3 showing all work below.

Suppose that the terminal side of angle α lies in Quadrant IV and the terminal side of angle β lies in Quadrant III. If $\cos\alpha = \dfrac{4}{7}$ and $\sin\beta = -\dfrac{8}{13}$, find the exact value of $\cos(\alpha+\beta)$.

Section 7.2 Objective 2 Understanding the Sum and Difference Formulas for the Sine Function

What are the two **Cofunction Identities for Sine and Cosine**?

What are the **Sum and Difference Formulas for the Sine Function**?

Work through the interactive video with Example 4 showing all work below.

Find the exact value of each trigonometric expression without the use of a calculator.

a. $\sin\left(-\dfrac{\pi}{3}+\dfrac{5\pi}{4}\right)$ b. $\sin(12°)\cos(78°)+\cos(12°)\sin(78°)$ c. $\sin(15°)$

308

Work through the video with Example 5 showing all work below.

Suppose that α is an angle such that $\tan\alpha = \dfrac{5}{7}$ and $\cos\alpha < 0$. Also, suppose that β is an angle such that $\sec\beta = -\dfrac{4}{3}$ and $\csc\beta > 0$. Find the exact value of $\sin(\alpha + \beta)$.

<u>Section 7.2 Objective 3 Understanding the Sum and Difference Formulas for the Tangent Function</u>

What are the two **Sum and Difference Formulas for the Tangent Function**?

Section 7.2

Work through the interactive video with Example 6 showing all work below.

Find the exact value of each trigonometric expression without the use of a calculator.

a. $\tan\left(\dfrac{5\pi}{6}+\dfrac{3\pi}{4}\right)$ b. $\tan\left(\dfrac{\pi}{12}\right)$

Section 7.2 Objective 4 Using the Sum and Difference Formulas to Verify Identities

Work through Example 7 and show all work below.

Verify the trigonometric identity $\sin(2\theta)=2\sin\theta\cos\theta$.

Work through the video with Example 8 and show all work below.

Verify the trigonometric identity $\csc(\alpha - \beta) = \dfrac{\sin \alpha \cos \beta + \cos \alpha \sin \beta}{\sin^2 \alpha - \sin^2 \beta}$.

<u>Section 7.2 Objective 5 Using the Sum and Difference Formulas to Evaluate Expressions Involving Inverse Trigonometric Functions</u>

Work through the video with Example 9 and show all work below.

Find the exact value of the expression $\cos(\sin^{-1}(\tfrac{1}{5}) + \cos^{-1}(-\tfrac{3}{4}))$ without using a calculator.

Section 7.3 Guided Notebook

7.3 The Double-Angle and Half-Angle Formulas
      Work through Section 7.3 TTK #1-6
      Work through Section 7.3 Objective 1
      Work through Section 7.3 Objective 2
      Work through Section 7.3 Objective 3
      Work through Section 7.3 Objective 4
      Work through Section 7.3 Objective 5

Section 7.3 The Double-Angle and Half-Angle Formulas

<u>7.3 Things To Know</u>

1. Evaluating Trigonometric Functions of Angles Belonging to the $\frac{\pi}{3}$, $\frac{\pi}{6}$, or $\frac{\pi}{4}$ Families (Section 5.4)
Try working through a "You Try It" problem or refer to Section 5.4 or watch the interactive video.

2. Finding the Exact and Approximate Values of an Inverse Sine Expression (Section 6.4)
Try working through a "You Try It" problem or refer to Section 6.4 or watch the interactive video.

3. Finding the Exact and Approximate Values of an Inverse Cosine Expression (Section 6.4)
Try working through a "You Try It" problem or refer to Section 6.4 or watch the interactive video.

4. Finding the Exact and Approximate Values of an Inverse Tangent Expression (Section 6.4)
Try working through a "You Try It" problem or refer to Section 6.4 or watch the interactive video.

5. Understanding the Sum and Difference Formulas for the Sine Function (Section 7.2)
Try working through a "You Try It" problem or refer to Section 7.2 or watch the interactive video.

6. Understanding the Sum and Difference Formulas for the Cosine Function (Section 7.2)
Try working through a "You Try It" problem or refer to Section 7.2 or watch the interactive video.

Section 7.3 Objective 1 Understanding the Double-Angle Formulas

What are the three **Double-Angle Formulas**?

What are the three **Double-Angle Formulas for Cosine**?

Work through the interactive video with Example 1 showing all work below.
  Rewrite each expression as the sine, cosine, or tangent of a double angle. Then evaluate the expression without using a calculator.

  a. $\cos^2(\frac{11\pi}{12}) - \sin^2(\frac{11\pi}{12})$

  b. $2\sin 67.5° \cos 67.5°$

  c. $\dfrac{2\tan(-\frac{\pi}{8})}{1 - \tan^2(-\frac{\pi}{8})}$

  d. $2\cos^2 105° - 1$

Work through the interactive video with Example 2 showing all work below.

Suppose that the terminal side of an angle θ lies in Quadrant II such that $\sin\theta = \dfrac{5}{7}$.

Find the values of $\sin 2\theta$, $\cos 2\theta$, and $\tan 2\theta$.

Work through the video with Example 3 showing all work below.

If $\tan 2\theta = -\dfrac{24}{7}$ for $\dfrac{3\pi}{2} < 2\theta < 2\pi$, then find the values of $\sin\theta$, $\cos\theta$, and $\tan\theta$.

<u>Section 7.3 Objective 2 Understanding the Power Reduction Formulas</u>

What are the three **Power Reduction Formulas**?

Work through the video with Example 4 showing all work below.
   Rewrite the function $f(x) = 6\sin^4 x$ as an equivalent function containing only cosine terms raised to a power of 1.

<u>Section 7.3 Objective 3 Understanding the Half-Angle Formulas</u>

What are **The Half-Angle Formulas for Sine and Cosine**?

315

What are **the Half-Angle Formulas for Tangent**?

Work through Example 5 showing all work below.
Use a half-angle formulas to evaluate each expression without using a calculator.

a. $\sin(-\frac{7\pi}{8})$

b. $\cos(-15°)$

c. $\tan(\frac{11\pi}{12})$

Work through the interactive video with Example 6 showing all work below.

Suppose that $\csc\alpha = \dfrac{8}{3}$ such that $\dfrac{\pi}{2} < \alpha < \pi$. Find the values of $\sin(\dfrac{\alpha}{2})$, $\cos(\dfrac{\alpha}{2})$, and $\tan(\dfrac{\alpha}{2})$.

<u>Section 7.3 Objective 4 Using the Double-Angle, Power Reduction, and Half-Angle Formulas to Verify Identities</u>

Work through Example 7 and show all work below.

Verify the trigonometric identity $\dfrac{\cos(2\theta)}{1+\sin(2\theta)} = \dfrac{\cos\theta - \sin\theta}{\cos\theta + \sin\theta}$.

Section 7.3 Objective 5 Using the Double-Angle and Half-Angle Formulas to Evaluate Expressions Involving Trigonometric Functions

Work through the video with Example 8 and show all work below.

Find the exact value of the expression $\cos(\frac{1}{2}\sin^{-1}(-\frac{3}{11}))$ without the use of a calculator.

Section 7.4 Guided Notebook

7.4 The Product-to-Sum and Sum-to-Product Formulas
Work through Section 7.4 TTK #1-3
Work through Section 7.4 Objective 1
Work through Section 7.4 Objective 2
Work through Section 7.4 Objective 3

Section 7.4 The Double-Angle and Half-Angle Formulas

7.4 Things To Know

1. Evaluating Trigonometric Functions of Angles Belonging to the $\frac{\pi}{3}$, $\frac{\pi}{6}$, or $\frac{\pi}{4}$ Families (Section 5.4)
Try working through a "You Try It" problem or refer to Section 5.4 or watch the interactive video.

2. Understanding the Sum and Difference Formulas for the Sine Function (Section 7.2)
Try working through a "You Try It" problem or refer to Section 7.2 or watch the interactive video.

3. Understanding the Sum and Difference Formulas for the Cosine Function (Section 7.2)
Try working through a "You Try It" problem or refer to Section 7.2 or watch the interactive video.

Section 7.4 Objective 1 Understanding the Product-to-Sum Formulas

What are the four **Product-to-Sum Formulas**?

319

Section 7.4

Work through the interactive video with Example 1 showing all work below.
Write each product as a sum or difference containing only sines or cosines.
a. $\sin 4\theta \sin 2\theta$

b. $\cos(\dfrac{19\theta}{2})\sin(\dfrac{\theta}{2})$

c. $\cos 11\theta \cos 5\theta$

d. $\sin 6\theta \cos 3\theta$

Work through the video with Example 2 showing all work below.

Determine the exact value of the expression $\sin(\frac{3\pi}{8})\cos(\frac{\pi}{8})$ without the use of a calculator.

Section 7.4 Objective 2 Understanding the Sum-to-Product Formulas

What are the four **Sum-to-Product Formulas**?

Work through the interactive video with Example 3 showing all work below.
   Write each sum or difference as a product of sines and/or cosines.
   a. $\sin 5\theta + \sin 3\theta$

   b. $\cos(\frac{3\theta}{2}) - \cos(\frac{17\theta}{2})$

Section 7.4

Work through the interactive video with Example 4 showing all work below.

Determine the exact value of the expression $\sin(\frac{\pi}{12}) - \sin(\frac{17\pi}{12})$ without the use of a calculator.

Section 7.4 Objective 3 Using the Product-to-Sum and Sum-to-Product Formulas to Verify Identities

Work through Example 5 showing all work below.

Verify the trigonometric identity $\dfrac{\cos\theta + \cos 3\theta}{2\cos 2\theta} = \cos\theta$.

322

Section 7.5 Guided Notebook

7.5   Trigonometric Equations
Work through Section 7.5 TTK #1-4
Work through Section 7.5 Objective 1
Work through Section 7.5 Objective 2
Work through Section 7.5 Objective 3
Work through Section 7.5 Objective 4
Work through Section 7.5 Objective 5

Section 7.5 Trigonometric Equations

7.5 Things To Know

1. Solving Equations That Are Quadratic in Form (Appendix B.6)
Try working through a "You Try It" problem or refer to Appendix B.6or watch the interactive video.

2. Evaluating Trigonometric Functions of Angles Belonging to the $\frac{\pi}{3}$, $\frac{\pi}{6}$, or $\frac{\pi}{4}$ Families (Section 5.4)
Try working through a "You Try It" problem or refer to Section 5.4 or watch the interactive video.

3. Finding the Exact and Approximate Values of an Inverse Sine Expression (Section 6.4)
Try working through a "You Try It" problem or refer to Section 6.4 or watch the interactive video.

4. Finding the Exact and Approximate Values of an Inverse Cosine Expression (Section 6.4)
Try working through a "You Try It" problem or refer to Section 6.4 or watch the interactive video.

Section 7.5 Introduction

What is the difference between **identities** and **conditional trigonometric equations**?

Section 7.5

What is the difference between **general solution(s)** and **specific solution(s)**?

Section 7.5 Objective 1 Solving Trigonometric Equations That Are Linear in Form

What are the four **Steps for Solving Trigonometric Equations That Are Linear in Form**?

Step 1.

Step 2.

Step 3.

Step 4.

Work through the interactive video with Example 1 showing all work below.
   Determine a general formula (or formulas) for all solutions to each equation. Then,
   determine the specific solutions (if any) on the interval $[0, 2\pi)$.

   a. $\sin \theta = \dfrac{1}{2}$

b. $\sqrt{3}\tan\theta+1=0$

c. $\sec\theta=-1$

Work through the interactive video with Example 2 showing all work below.
   Determine a general formula (or formulas) for all solutions to each equation. Then, determine the specific solutions (if any) on the interval $[0,2\pi)$.

a. $\sqrt{2}\cos 2\theta+1=0$

b. $\sin\dfrac{\theta}{2}=-\dfrac{\sqrt{3}}{2}$

c. $\tan(\theta+\dfrac{\pi}{6})+1=0$

Section 7.5 Objective 2 Solving Trigonometric Equations That Are Quadratic in Form

Work through the interactive video with Example 3 showing all work below.
  Determine a general formula (or formulas) for all solutions to each equation. Then, determine the specific solutions (if any) on the interval $[0, 2\pi)$.

  a. $\sin^2 \theta - 4\sin \theta + 3 = 0$

  b. $4\cos^2 \theta - 3 = 0$

Section 7.5 Objective 3 Solving Trigonometric Equations Using Identities

Work through the interactive video with Example 4 showing all work below.
  Determine a general formula (or formulas) for all solutions to each equation. Then, determine the specific solutions (if any) on the interval $[0, 2\pi)$.

  a. $2\sin^2 \theta = 3\cos \theta + 3$

b. $\sin\theta\cos\theta = -\dfrac{1}{2}$

c. $\cos 2\theta + 4\sin^2\theta = 2$

d. $\sin 5\theta + \sin 3\theta = 0$

Section 7.5 Objective 4 Solving Other Types of Trigonometric Equations

Work through the interactive video with Example 5 and show all work below.

Determine a general formula (or formulas) for all solutions to each equation. Then, determine the specific solutions (if any) on the interval $[0, 2\pi)$.

a. $\sin 2\theta + 2\cos \theta \sin 2\theta = 0$

b. $\cos^2 \theta = \sin \theta \cos \theta$

c. $\sin \theta + \cos \theta = 1$

Section 7.5 Objective 5 Solving Trigonometric Equations Using a Calculator

Work through Example 6 and show all work below.

Approximate all solutions to the trigonometric equation $\cos\theta = -0.4$ on the interval $[0, 2\pi)$. Round your answer to four decimal places.

Why can't we simply use an inverse trigonometric function key on our calculator to solve this trigonometric equation? (Hint: See the Caution text on page 7.5-46.)

Section 8.1 Guided Notebook

8.1   Right Triangle Trigonometry
Work through Section 8.1 TTK #1
Work through Section 8.1 Objective 1

Section 8.1 Right Triangle Trigonometry

<u>**8.1 Things To Know**</u>

1. Understanding the Right Triangle Definitions of the Trigonometric Functions (Section 5.4)

Try working through a "You Try It" problem or refer to Section 5.4or watch the video.

<u>Section 8.1 Introduction</u>

Write down the Right Triangle Definitions of the six Trigonometric Functions

Try working through the two "You Try IT" examples on page 8.1-3 to practice using a calculator to evaluate trigonometric expressions given in degrees or radians.

Section 8.1

Section 8.1 Objective 1 Applications of Right Triangle Trigonometry

Work through Example 1 and show all work below.
Suppose that the length of the hypotenuse of a right triangle is 11 inches. If one of the acute angles is $37°$, find the length of the two legs. Round to two decimal places.

What is the **angle of elevation**? What is the **angle of depression**? Sketch and label the diagram shown in Figure 1.

Work through the video with Example 2 and show all work below.

The angle of elevation to the top of a flagpole measured by a digital protractor is $\dfrac{\pi}{9}$ radians from a point on the ground 90 feet away from its base. Find the height of the flagpole. Round to two decimal places.

Work through the video with Example 3 and show all work below.

At the same instant, two observers 5 miles apart are looking at the same airplane. (See Figure 37.) The angle of elevation (from the ground) of the observer closest to the plane is $71°$. The angle of elevation (from the ground) of the person furthest from the plane is $39°$. Find the altitude of the plane (to the nearest foot) at the instant the two people observe the plane.

Section 8.2 Guided Notebook

8.2   The Law of Sines
    Work through Section 8.2 TTK #1
    Work through Section 8.2 Objective 1
    Work through Section 8.2 Objective 2
    Work through Section 8.2 Objective 3
    Work through Section 8.2 Objective 4

Section 8.2 The Law of Sines

8.2 Things To Know

1. Understanding the Inverse Sine Function (Section 6.4)
Try working through a "You Try It" problem or refer to Section 6.4 or watch the interactive video.

Section 8.2 Objective 1 Determining If the Law of Sines Can be Used to Solve an Oblique Triangle

What are the two types of **oblique triangles**? (Describe and sketch.)

What is the **Law of Sines**?

Section 8.2

Describe the six cases of oblique triangles, as seen in Table 1.

What three pieces of information are needed to solve an oblique triangle using the Law of Sines?

1.

2.

3.

Work through the video with Example 1 showing all work below.
 Decide whether or not the Law of Sines can be used to solve each triangle. Do not attempt to solve the triangle. (See page 8.2-7 of eText for pictures of the triangles.)

Section 8.2 Objective 2 Using the Law of Sines to Solve the SAA Case or ASA Case

Work through Example 2 showing all work below.
Solve the given oblique triangle (see page 8.2-10 of eText). Round the lengths of the sides to one decimal place.

Work through the video with Example 3 showing all work below.
Solve the oblique triangle ABC (see page 8.2-13 of eText) if $B = 38°$, $C = 72°$, and $a = 7.5$ cm. Round the lengths of the sides to one decimal place.

Section 8.2 Objective 3 Using the Law of Sines to Solve the SSA Case

Describe the different scenarios for the SSA case as seen in Table 2.

Work through Example 4 and show all work below.

Two sides and an angle are given below. Determine whether the information results in no triangle, one right triangle, or one or two oblique triangles. Solve each resulting triangle. Round the measures of all angles and the lengths of all sides to one decimal place.

$$a = 10 \text{ ft}, \quad b = 28 \text{ ft}, \quad A = 29°$$

Work through the video with Example 5 and show all work below.

Two sides and an angle are given below. Determine whether the information results in no triangle, one right triangle, or one or two oblique triangles. Solve each resulting triangle. Round the measures of all angles and the lengths of all sides to one decimal place.

$$a = 13 \text{ cm}, \quad b = 7.8 \text{ cm}, \quad A = 67°$$

Work through the video with Example 6 and show all work below.

Two sides and an angle are given below. Determine whether the information results in no triangle, one right triangle, or one or two oblique triangles. Solve each resulting triangle. Round the measures of all angles and the lengths of all sides to one decimal place.

$$b = 11.3 \text{ in.}, \quad c = 15.5 \text{ in.}, \quad B = 34.7°$$

Section 8.2

What are the three cases for which the Law of Sines can be used? (Sketch and label.)

Section 8.2 Objective 4 Using the Law of Sines to Solve Applied Problems Involving Oblique Triangles

Work through Example 7 and show all work below.

> To determine the width of a river, forestry workers place markers on opposite sides of the river at points A and B. A third marker is placed at point C, 200 feet away from point A forming triangle ABC. If the angle in triangle ABC at point C is 51° and if the angle in triangle ABC at point A is 110°, then determine the width of the river rounded to the nearest tenth of a foot. (See the image on page 8.2-25.)

Describe the concept of **bearing**.

Work through the video with Example 8 and show all work below.

    A ship set sail from port at a bearing of N 53° E and sailed 63 km to point *B*. The ship then turned and sailed an additional 69 km to point *C*. Determine the distance from port to point *C* if the ship's final bearing is N 74° E. Round to the nearest tenth of a kilometer.

Section 8.3 Guided Notebook

8.3   The Law of Cosines
Work through Section 8.3 TTK #1-2
Work through Section 8.3 Objective 1
Work through Section 8.3 Objective 2
Work through Section 8.3 Objective 3
Work through Section 8.3 Objective 4

Section 8.3 The Law of Cosines

<u>**8.3 Things To Know**</u>

1. Understanding the Inverse Sine Function (Section 6.4)
Try working through a "You Try It" problem or refer to Section 6.4 or watch the interactive video.

2. Understanding the Inverse Cosine Function (Section 6.4)
Try working through a "You Try It" problem or refer to Section 6.4 or watch the interactive video.

<u>Section 8.3 Objective 1 Determining If the Law of Sines or the Law of Cosines Should Be Used to Begin to Solve an Oblique Triangle</u>

What is the Law **of Cosines**?

What is the **Alternate Form of the Law of Cosines**?

Work through the video with Example 1 showing all work below.

Decide whether the Law of Sines or the Law of Cosines should be used to begin to solve the given triangle (see page 8.3-8 in eText). Do not solve the triangle.

Section 8.3 Objective 2 Using the Law of Cosines to Solve the SAS Case

What are the three steps for **Solving an SAS Oblique Triangle**?

Work through the interactive video with Example 2 showing all work below.

Solve the given oblique triangle (see page 8.3-11 of eText). Round the measures of all angles and the lengths of all sides to one decimal place.

Section 8.3

Section 8.3 Objective 3 Using the Law of Cosines to Solve the SSS Case

What are the three steps for **Solving a SSS Oblique Triangle**?

Work through the interactive video with Example 3 and show all work below.
Solve oblique triangle ABC if $a = 5$ ft, $b = 8$ ft, and $c = 12$ ft.

Section 8.3 Objective 4 Using the Law of Cosines to Solve Applied Problems Involving Oblique Triangles

Work through the video with Example 4 and show all work below.
Two planes take off from different runways at the same time. One plane flies at an average speed of 350 mph with a bearing of N 21° E. The other plane flies at an average speed of 420 mph with a bearing of S 84° W. How far are the planes from each other 2 hours after takeoff? Round to the nearest tenth of a mile.

Work through Example 5 and show all work below.

A **chord** of a circle is a line segment with endpoints that both lie on the circumference of a circle. Determine the measure of the central angle (in degrees) if the length of the chord intercepted by the central angle of a circle of radius 10 inches is 16.5 inches. Round the measure of the central angle to one decimal place.

Section 8.4 Guided Notebook

8.4   Area of Triangles
Work through Section 8.4 TTK #1-4
Work through Section 8.4 Objective 1
Work through Section 8.4 Objective 2
Work through Section 8.4 Objective 3

Section 8.4 Area of Triangles

8.4 Things To Know

1. Understanding the Inverse Sine Function (Section 6.4)
Try working through a "You Try It" problem or refer to Section 6.4 or watch the interactive video.

2. Understanding the Inverse Cosine Function (Section 6.4)
Try working through a "You Try It" problem or refer to Section 6.4 or watch the interactive video.

3. Using the Law of Sines to Solve the SAA Case or the ASA Case (Section 8.2)
Try working through a "You Try It" problem or refer to Section 8.2 or watch the video.

4. Using the Law of Cosines to Solve the SSS Case (Section 8.2)
Try working through a "You Try It" problem or refer to Section 8.2 or watch the video.

Section 8.4 Objective 1 Determining the Area of Oblique Triangles

How do we compute the **Area of a Triangle** in terms of the length of the base and the height?

How do we compute the **Area of a Triangle** in terms of angles? (Hint: See the text box at the top of page 8.4-4.)

Work through the video with Example 1 showing all work below.

Determine the area of each triangle (see page 8.4-4 of eText). Round each answer to two decimal places.

Section 8.4 Objective 2 Using Heron's Formula to Determine the Area of an SSS Triangle

Work through Example 2 showing all work below.

Determine the area of the given triangle (see page 8.4-8 of eText). Round to two decimal places.

Section 8.4

What is **Heron's formula**?

Work through the video with Example 3 showing all work below.

    Use Heron's formula to determine the area of the given triangle (see page 8.4-11).
    Round to two decimal places.

Section 8.4 Objective 3 Solving Applied Problems Involving the Area of Triangles

Work through Example 4 and show all work below.

    A painter who is painting a house has only one side of the house left to paint. He has
    enough paint to cover 1200 square feet. A cross section of the unpainted side of the
    house is shown in the figure (see page 8.4-13). What is the area of the unpainted side
    of the house? Does he have enough paint to finish the job?

Section 9.1 Guided Notebook

9.1 Polar Coordinates and Polar Equations
  Work through Section 9.1 TTK #1-5
  Work through Section 9.1 Objective 1
  Work through Section 9.1 Objective 2
  Work through Section 9.1 Objective 3
  Work through Section 9.1 Objective 4
  Work through Section 9.1 Objective 5
  Work through Section 9.1 Objective 6

Section 9.1 Polar Coordinates and Polar Equations

9.1 Things To Know

1. Understanding the Four Families of Special Angles (Section 5.4)
Try working through a "You Try It" problem or refer to Section 5.4 or watch the interactive video.

2. Understanding the Definitions of the Trigonometric Functions of General Angles (Section 5.4)
Try working through a "You Try It" problem or refer to Section 5.4 or watch the video.

3. Understanding the Signs of the Trigonometric Functions (Section 5.4)
Try working through a "You Try It" problem or refer to Section 5.4 or watch the video.

4. Evaluating Trigonometric Functions of Angles Belonging to the $\frac{\pi}{3}$, $\frac{\pi}{6}$, or $\frac{\pi}{4}$ Families

(Section 5.4)
Try working through a "You Try It" problem or refer to Section 5.4 or watch the interactive video.

5. Solving Trigonometric Equations That Are Linear in Form (Section 7.5)
Try working through a "You Try It" problem or refer to Section 7.5 or watch the interactive video.

Section 9.1

Section 9.1 Objective 1 Plotting Points Using Polar Coordinates

What is the **polar coordinate system**?

What is the definition of **Directed Distance**?

Sketch and label a polar grid, as seen in Figure 2.

Work through the video with Example 1 showing all work below.
   Plot the following points in a polar coordinate system.

   a. $A(3, \frac{\pi}{4})$ b. $B(-2, 120°)$ c. $C(1.5, -\frac{7\pi}{6})$ d. $D(-3, -\frac{3\pi}{4})$

Section 9.1 Objective 2 Determining Different REpresetations of the Point (r, θ)

What are the two ways to obtain different representations of the same point?

Work through the video with Example 2 showing all work below.

The point $P(4, \frac{5\pi}{6})$ is shown in Figure 7 (see page 9.1-12 of eText). Determine three different representations of point P that have the specified conditions.

a. $r > 0, \ -2\pi \le \theta < 0$ b. $r < 0, \ 0 \le \theta < 2\pi$ c. $r > 0, \ 2\pi \le \theta < 4\pi$

Section 9.1 Objective 3 Converting a Point from Polar Coordinates to Rectangular Coordinates

What are the **Relationships Used when Converting a Point from Polar Coordinates to Rectangular Coordinates**?

Section 9.1

Work through the video with Example 3 and show all work below.
    Determine the rectangular coordinates for the points with the given polar coordinates.

    a. $A(5, \pi)$ b. $B(-7, -\frac{\pi}{3})$ c. $C(3\sqrt{2}, \frac{5\pi}{4})$

Section 9.1 Objective 4 Converting a Point from Rectangular Coordinates to Polar Coordinates

For simplicity and consistency, we will always determine the polar

coordinates with the conditions that _____ and _____.

Work through the video with Example 4 and show all work below.
    Determine the polar coordinates for the points with the given rectangular coordinates.

    a. $A(-3.5, 0)$ b. $B(0, -\sqrt{7})$

What are the four steps for **Converting Rectangular Coordinates to Polar Coordinates for Points Not Lying Along an Axis**?

Step 1.

Step 2.

Step 3.

Step 4.

Work through the interactive video with Example 5 and show all work below.

Determine the polar coordinates for the points with the given rectangular coordinates such that $r \geq 0$ and $0 \leq \theta < 2\pi$.

a. $A(-4,-4)$ b. $B(-2\sqrt{3},2)$ c. $C(4,-3)$

Section 9.1 Objective 5 Converting an Equation from Rectangular Form to Polar Form

What is a polar equation? (Define and give examples.)

Work through the video with Example 6 and show all work below.

Convert each equation given in rectangular form into polar form.

a. $x = 7$

b. $2x - y = 8$

c. $4x^2 + 4y^2 = 3$

d. $x^2 + y^2 = 9y$

Section 9.1 Objective 6 Converting an Equation from Polar Form to Rectangular Form

Work through the video with Example 7 and show all work below.
Convert each equation given in polar form into rectangular form.
a. $3r\cos\theta - 4r\sin\pi = -1$

b. $r = 6\cos\pi$

c. $r = 3$

d. $\theta = \dfrac{\pi}{6}$

Section 9.2 Guided Notebook

9.2   Graphing Polar Equations
Work through Section 9.2 TTK #1-4
Work through Section 9.2 Objective 1
Work through Section 9.2 Objective 2
Work through Section 9.2 Objective 3
Work through Section 9.2 Objective 4
Work through Section 9.2 Objective 5

Section 9.2 Graphing Polar Equations

9.2 Things To Know

1. Evaluating Trigonometric Functions of Angles Belonging to the $\dfrac{\pi}{3}$, $\dfrac{\pi}{6}$, or $\dfrac{\pi}{4}$ Families (Section 5.4)
Try working through a "You Try It" problem or refer to Section 5.4 or watch the interactive video.

2. Solving Trigonometric Equations That Are Linear in Form (Section 7.5)
Try working through a "You Try It" problem or refer to Section 7.5 or watch the interactive video.

3. Plotting Points Using Polar Coordinates (Section 9.1)
Try working through a "You Try It" problem or refer to Section 9.1 or watch the video.

4. Converting an Equation from Polar Form to Rectangular Form (Section 9.1)
Try working through a "You Try It" problem or refer to Section 9.1 or watch the video.

Section 9.2 Objective 1 Sketching Equations of the Form
$\theta = \alpha,\ r\cos\theta = a,\ r\sin\theta = a\ \ and\ \ ar\cos\theta + br\sin\theta = c$

Work through the video with Example 1 showing all work below.

Sketch the graph of the polar equation $\theta = \dfrac{2\pi}{3}$.

Work through Example 2 showing all work below.
Sketch the graph of the polar equation $r\cos\theta = 2$.

Work through the video with Example 3 showing all work below.
Sketch the graph of the polar equation $r\sin\theta = -3$.

357

Section 9.2

Work through Example 4 showing all work below.

Sketch the graph of the polar equation $3r \cos \pi - 2r \sin \pi = -6$

Write down the summary, **Graphs of Polar Equations of the Form**
$\theta = \alpha$, $r \cos \theta = a$, $r \sin \theta = a$, **and** $ar \cos \theta + br \sin \theta = c$ **, where a, b, and c are Constants**
found on pages 9.2-11 and 9.2-12.

Section 9.2 Objective 2 Sketching Equations of the Form
$r = a$, $r = a \sin \pi$, and $r = a \cos \pi$

Work through Example 5 showing all work below.
Sketch the graph of the polar equation $r = -3$.

Work through the interactive video with Example 6 showing all work below.
Sketch the graph of each polar equation.
a. $r = 4 \sin \theta$

b. $r = -2 \cos \theta$

Section 9.2

Write down the summary, **Graphs of Polar Equations of the Form** $r = a$, $r = a\sin\theta$, **and** $r = a\cos\theta$, **where** $a \neq 0$ **is a Constant** found on page 9.2-17.

Section 9.2 Objective 3 Sketching Equations of the Form
$r = a + b\sin\theta$ and $r = a + b\cos\theta$

What are **limacons**? What form do their equations have?

Work through Example 7 showing all work below.
    Sketch the graph of the polar equation $r = 3 - 3\sin\theta$.

360

What is a **cardioid**?

Work through Example 8 showing all work below.
Sketch the graph of the polar equation $r = -1 + 2\cos\theta$.

Work through Example 9 showing all work below.
Sketch the graph of the polar equation $r = 3 + 2\sin\theta$..

361

Write down the summary, **Graphs of Polar Equations of the Form** $r = a + b\sin\theta$, **and** $r = a + b\cos\theta$, **where** $a \neq 0$ **and** $b \neq 0$ **Are Constants** found on pages 9.2-27 and 9.2-28.

What are the four **Steps for Sketching Polar Equations (Limacons) of the Form**
$r = a + b\sin\theta$, **and** $r = a + b\cos\theta$.

Step 1.

Step 2.

Step 3.

Step 4.

Work through the interactive video with Example 10 showing all work below.
Sketch the graph of each polar equation.
a. $r = 4 - 3\cos\theta$

b. $r = 2 + \sin \theta$

c. $r = -2 + 2 \cos \theta$

d. $r = 3 - 4 \sin \theta$

Section 9.2 Objective 4 Sketching Equations of the Form
$r = a \sin n\theta$ and $r = a \cos n\theta$

What do we call the graphs of these types of equations? Why?

Work through Example 11 and show all work below.
    Sketch the graph of $r = 3 \sin 2\theta$.

If n is **even**, there will be _____ petals.

If n is **odd**, there will be _____ petals.

Section 9.2

Write down the summary, **Graphs of Polar Equations of the Form** $r = a\sin n\theta$ **, and** $r = a\cos n\theta$ **, where** $a \neq 0$ **Is a Constant and** $n \neq 1$ **Is a Positive Integer** found on pages 9.2-36 and 9.2-37.

What are the six **Steps for Sketching Polar Equations (Roses) of the Form** $r = a\sin n\theta$ **, and** $r = a\cos n\theta$ **, where** $a \neq 0$ **Is a Constant and** $n \neq 1$ **Is a Positive Integer**

      Step 1.

      Step 2.

      Step 3.

      Step 4.

      Step 5.

      Step 6.

Work through the interactive video with Example 12 showing all work below.

Sketch the graph of each polar equation.

a. $r = -4\cos 3\theta$

b. $r = -2\sin 5\theta$

c. $r = 5\cos 4\theta$

<u>Section 9.2 Objective 5 Sketching Equations of the Form</u>
<u>$r^2 = a^2 \sin 2\theta$ and $r^2 = a^2 \cos 2\theta$</u>

What is a **lemniscate**? What form do their equations have?

Write down the summary, **Graphs of Polar Equations of the Form**
$r^2 = a^2 \sin 2\theta$ and $r^2 = a^2 \cos 2\theta$, **where** $a \neq 0$ **Is a Constant** found on page 9.2-44.

Work through the video with Example 13 showing all work below.
Sketch the graph of each polar equation.
a. $r^2 = 9 \cos 2\theta$

b. $r^2 = 16 \sin 2\theta$

Section 9.3 Guided Notebook

9.3   Complex Numbers in Polar Form; De Moivre's Theorem

  Work through Section 9.3 TTK #1-4
  Work through Section 9.3 Objective 1
  Work through Section 9.3 Objective 2
  Work through Section 9.3 Objective 3
  Work through Section 9.3 Objective 4
  Work through Section 9.3 Objective 5
  Work through Section 9.3 Objective 6
  Work through Section 9.3 Objective 7

Section 9.3 Complex Numbers in Polar Form; De Moivre's Theorem

9.3 Things To Know

1. Understanding the Four Families of Special Angles (Section 5.4)
Try working through a "You Try It" problem or refer to Section 5.4 or watch the interactive video.

2. Understanding the Definitions of the Trigonometric Functions of General Angles (Section 5.4)
Try working through a "You Try It" problem or refer to Section 5.4 or watch the video.

3. Evaluating Trigonometric Functions of Angles Belonging to the $\dfrac{\pi}{3}$, $\dfrac{\pi}{6}$, or $\dfrac{\pi}{4}$ Families
(Section 5.4)
Try working through a "You Try It" problem or refer to Section 5.4 or watch the interactive video.

4. Solving Trigonometric Equations That Are Linear in Form (Section 7.5)
Try working through a "You Try It" problem or refer to Section 7.5 or watch the interactive video.

5. Plotting Points Using Polar Coordinates (Section 9.1)
Try working through a "You Try It" problem or refer to Section 9.1 or watch the video.

6. Converting a Point from Rectangular Coordinates to Polar Coordinates (Section 9.1)
Try working through a "You Try It" problem or refer to Section 9.1 or watch the interactive video.

Section 9.3

<u>Section 9.3 Objective 1 Understanding the Rectangular Form of a Complex Number</u>

What is the definition of **The Rectangular Form of a Complex Number**?

What is the **complex plane**? (Define and sketch.)

What is the definition of **The Absolute Value of a Complex Number**?

Work through the video with Example 1 showing all work below.
    Plot each complex number in the complex plane and determine its absolute value.
    a. $z_1 = 3 - 4i$ b. $z_2 = -2 + 5i$ c. $z_3 = 3$ d. $z_4 = -2i$

Section 9.3 Objective 2 Understanding the Polar Form of a Complex Number

What is the definition of **The Polar Form of a Complex Number**?

What is the **modulus**? What is the **argument**?

Complex numbers in polar form, where $0 \le \theta < 2\pi$ (or $0 \le \theta < 360°$) are said

to be in _____.

Work through the video with Example 2 showing all work below.
    Rewrite the complex number in standard polar form, plot the number in the complex plane, and determine the quadrant in which the point lies or the axis on which the point lies.

    a. $z = 3(\cos\dfrac{5\pi}{8} + i\sin\dfrac{5\pi}{8})$

    b. $z = 2(\cos\dfrac{23\pi}{4} + i\sin\dfrac{23\pi}{4})$

    c. $z = 4(\cos(-3\pi) + i\sin(-3\pi))$

Section 9.3 Objective 3 Converting a Complex Number from Polar Form to Rectangular Form

Work through the video with Example 3 showing all work below.

      Write each complex number in rectangular form using exact values if possible. Otherwise, round to two decimal places.

     a. $z = 3(\cos\dfrac{7\pi}{4} + i\sin\dfrac{7\pi}{4})$

     b. $z = 4(\cos 80° + i\sin 80°)$

Section 9.3 Objective 4 Converting a Complex Number from Rectangular Form to Standard Polar Form

Write down the four cases outlined in **Converting Complex Numbers From Rectangular Form to Standard Polar Form for Complex Numbers Lying Along the Real Axis or Imaginary Axis**.

Work through the video with Example 4 showing all work below.

Determine the standard polar form of each complex number. Write the argument using radians.

a. $z = 5$

b. $z = -3i$

c. $z = -\sqrt{7}$

d. $z = \dfrac{7}{2}i$

373

Section 9.3

What are the four steps for **Converting a Complex Number from Rectangular Form to Standard Polar Form for** $a \neq 0$ **and** $b \neq 0$?

Step 1.

Step 2.

Step 3.

Step 4.

Work through the interactive video with Example 5 showing all work below.

Determine the standard polar form of each complex number. Write the argument in radians using exact values if possible. Otherwise, round the argument to two decimal places.

a. $z = -2\sqrt{3} + 2i$

b. $z = 4 - 3i$

Section 9.3 Objective 5 Determining the Product or Quotient of Complex Numbers in Polar Form

How do we compute **The Product and Quotient of Two Complex Numbers Written in Polar Form**?

Work through the video with Example 6 showing all work below.

Let $z_1 = 4(\cos\frac{2\pi}{3} + i\sin\frac{2\pi}{3})$ and $z_2 = 5(\cos\frac{11\pi}{6} + i\sin\frac{11\pi}{6})$. Find $z_1 z_2$ and $\frac{z_1}{z_2}$ and write the answers in standard polar form.

Section 9.3

Section 9.3 Objective 6 Using De Moivre's Theorem to Raise a Complex Number to a Power

What is **De Moivre's Theorem for Finding Powers of a Complex Number**?

Work through the interactive video with Example 7 showing all work below.

a. Find $[5(\cos\frac{3\pi}{4} + i\sin\frac{3\pi}{4})]^3$ and write your answer in standard polar form.

b. Find $(\sqrt{3} - i)^4$ and write your answer in rectangular form.

Section 9.3 Objective 7 Using De Moivre's Theorem to Find the Roots of a Complex Number

What is **De Moivre's Theorem for Finding the *n*th Roots of a Complex Number**?

Work through Example 8 showing all work below.

Find the complex cube roots of 8. Write your answers in rectangular form.

Section 9.3

If we can find z_0, then we can easily determine the remaining n-1 roots by

adding _____ to each successive argument.

Work through the video with Example 9 showing all work below.

      a. Find the complex fourth roots of $z = 81(\cos\frac{3\pi}{5} + i\sin\frac{3\pi}{5})$. Write your answers in

      standard polar form.

      b. Find the complex square roots of $z = -2\sqrt{3} + 2i$. Write your answer in rectangular
      form with each part rounded to two decimal places.

Section 9.4 Guided Notebook

9.4 Vectors

Work through Section 9.4 TTK #1-3
Work through Section 9.4 Objective 1
Work through Section 9.4 Objective 2
Work through Section 9.4 Objective 3
Work through Section 9.4 Objective 4
Work through Section 9.4 Objective 5
Work through Section 9.4 Objective 6
Work through Section 9.4 Objective 7
Work through Section 9.4 Objective 8
Work through Section 9.4 Objective 9

Section 9.4 Vectors

9.4 Things To Know

1. Converting a Point from Polar Coordinates to Rectangular Coordinates (Section 9.1)
Try working through a "You Try It" problem or refer to Section 9.1 or watch the video.

2. Converting a Point from Rectangular Coordinates to Polar Coordinates (Section 9.1)
Try working through a "You Try It" problem or refer to Section 9.1 or watch the interactive video.

3. Converting a Complex Number from Rectangular Form to Standard Polar Form (Section 9.3)
Try working through a "You Try It" problem or refer to Section 9.3 or watch the interactive video.

379

Section 9.4

Section 9.4 Introduction

What is a **vector**?

Section 9.4 Objective 1 Understanding the Geometric Representation of a Vector

What is the **initial point**? What is the **terminal point**?

What is the definition of the **Geometric Representation of a Vector**?

What is the **zero vector**?

What is the definition of **The Magnitude of a Vector Represented Geometrically**?

Work through the video with Example 1 showing all work below.
    Determine the magnitude of **v**. (See page 9.4-6 of the eText.)

380

Section 9.4 Objective 2 Understanding Operations on Vector Represented Geometrically

In **scalar multiplication**, what happens when k is greater than zero? What happens when k is less than zero?

Work through Example 2 showing all work below.

    Given the vector **v** (see page 9.4-8 of eText), draw the vectors $2\mathbf{v}$ and $-\dfrac{1}{2}\mathbf{v}$.

What is the definition of **Parallel Vectors**?

In **Vector Addition**, what is the **triangle law**?

In **Vector Addition**, what is the **parallelogram law**?

Section 9.4

Work through the video with Example 3 showing all work below.
  Given the vectors **u, v,** and **w** (see page 9.4-12 of eText), draw each of the following
  vectors.
  a. **w + v**   b. **v − u**   c. **v + 2w − u**

Section 9.4 Objective 3 Understanding Vectors in Terms of Components

What is the definition of **Equal Vectors**?

What is the definition of **Representing a Vector in Terms of Components** *a* **and** *b***?**

Work through Example 4 and show all work below.
  Determine the component representation and the magnitude of a vector **v** having an
  initial point $P(5,3)$ and a terminal point $Q(-6,5)$.

What is the definition of **A Vector in Standard Position**?

Section 9.4 Objective 4 Understanding Vectors Represented in Terms of **i** and **j**

What is the definition of a **Unit Vector**?

What is the definition of the **Unit Vectors i and j**?

What is the definition of **A Vector Represented in Terms of i and j**?

Work through Example 5 and show all work below.
Write the vector $\mathbf{v} = \langle -5, 2 \rangle$ in terms of the unit vectors **i** and **j**.

What are the **Operations with Vectors in Terms of i and j**?

Work through the video with Example 6 and show all work below.

Let $\mathbf{u} = -3\mathbf{i} + 7\mathbf{j}$ and $\mathbf{v} = 5\mathbf{i} - \mathbf{j}$. Find each vector in terms of $\mathbf{i}$ and $\mathbf{j}$ and determine the magnitude of each vector.

a. $-\dfrac{1}{2}\mathbf{u}$ b. $\mathbf{u} + \mathbf{v}$ c. $\mathbf{u} - \mathbf{v}$ d. $3\mathbf{u} - 5\mathbf{v}$

What are the ten **Properties of Vectors**?

1.

2.

3.

4.

5.

6.

7.

8.

9.

10.

Section 9.4 Objective 5 Finding a Unit Vector

What is the definition of **The Unit Vector in the Same Direction of a Given Vector**?

Work through the video with Example 7 and show all work below.
Find the unit vector that has the same direction as **v** = 6**i** – 8**j**.

Section 9.4 Objective 6 Determining the Direction Angle of a Vector

What is the definition of **The Direction Angle of a Vector**?

Work through the video with Example 8 and show all work below.
Determine the direction angle of the vector **v** = -3**i** + 2**j**.

385

Section 9.4

Section 9.4 Objective 7 Representing a Vector in Terms of **i** and **j** Given Its Magnitude and Direction Angle

Describe the procedure for **Representing a Vector in Terms of i and j Given Its Magnitude and Direction Angle**.

Work through the video with Example 9 and show all work below.
   The vector **v** has a magnitude of 20 units and direction angle of $\theta = 50°$. Represent this vector in the form **v = a i + b j**. Round a and b to two decimal places.

Section 9.4 Objective 8 Using Vectors to Solve Applications Involving Velocity

Work through the video with Example 10 and show all work below.
   An airplane takes off from a runway at a speed of 190 mph at an angle of 11°. Express the velocity of the plane at takeoff as a vector in terms of **i** and **j**. Round a and b to two decimal places.

386

Work through the video with Example 11 and show all work below.

The wind is blowing at a speed of 35 mph in a direction of N 30° E. Express the velocity of the wind as a vector in terms of **i** and **j**.

Work through the video with Example 12 and show all work below.

A 747 jet was heading due east 520 mph in still air and encountered a 60 mph headwind blowing in the direction N 40° W. Determine the resulting ground speed of the plane and its new bearing. Round the ground speed to the nearest hundredth.

Section 9.4 Objective 9 Using Vectors to Solve Applications Involving Force

What is the definition of **Static Equilibrium**?

Work through the video with Example 13 and show all work below.

The forces $\mathbf{F}_1 = 6\mathbf{i} - 8\mathbf{j}$ and $\mathbf{F}_2 = 3\mathbf{i} + 2\mathbf{j}$ are acting on an object. What additional force is required for the object to be in static equilibrium?

Work through the video with Example 14 and show all work below.

Two tugboats are towing a large ship out of port and into the open sea. One tugboat exerts a force of $\| \mathbf{F}_1 \| = 2000$ pounds in a direction N 35° W. The other tugboat pulls with a force of $\| \mathbf{F}_2 \| = 1400$ pounds in a direction S 55° W. Find the magnitude of the resultant force and the bearing of the ship.

Section 9.5 Guided Notebook

9.5 The Dot Product

Work through Section 9.5 TTK #1-4
Work through Section 9.5 Objective 1
Work through Section 9.5 Objective 2
Work through Section 9.5 Objective 3
Work through Section 9.5 Objective 4
Work through Section 9.5 Objective 5
Work through Section 9.5 Objective 6

Section 9.5 The Dot Product

9.5 Things To Know

1. Understanding Vectors in Terms of **i** and **j** (Section 9.4)
Try working through a "You Try It" problem or refer to Section 9.4 or watch the video.

2. Finding a Unit Vector (Section 9.4)
Try working through a "You Try It" problem or refer to Section 9.4 or watch the video.

3. Determining the Direction Angle of a Vector (Section 9.4)
Try working through a "You Try It" problem or refer to Section 9.4 or watch the video.

4. Representing a Vector in Terms of **i** and **j** (Section 9.4)
Try working through a "You Try It" problem or refer to Section 9.4 or watch the video.

Section 9.5

Objective 1 Understanding the Dot Product and its Properties

What is the definition of **The Dot Product of Two Vectors**?

What are two other names for the dot product?

Work through Example 1 showing all work below.
    If $\mathbf{u} = -3\mathbf{i} + 5\mathbf{j}$ and $\mathbf{v} = 7\mathbf{i} - 4\mathbf{j}$, then find $\mathbf{u} \cdot \mathbf{v}$ and $\mathbf{v} \cdot \mathbf{u}$.

What are the five **Dot Product Properties**?

Work through the video with Example 2 showing all work below.
    If $\mathbf{u} = -4\mathbf{i} + 6\mathbf{j}$, $\mathbf{v} = -2\mathbf{i} + 8\mathbf{j}$, and $\mathbf{w} = -3\mathbf{i} - \mathbf{j}$ then find each of the following.
    a. $\mathbf{u} \cdot \mathbf{v}$ b. $\mathbf{u} \cdot (\mathbf{v} + \mathbf{w})$ c. $\mathbf{u} \cdot (-5\mathbf{v})$ d. $\| \mathbf{w} \|^2$

Section 9.5 Objective 2 Using the Dot Product to Determine the Angle between Two Vectors

If **u** and **v** are non-zero vectors and if θ is the angle between **u** and **v**, then

_____.

Work through the video with Example 3 showing all work below.

Determine the angle between each pair of vectors. Give the angle in degrees rounded to the nearest hundredth of a degree.

a. $\mathbf{u} = \mathbf{i} + 4\mathbf{j}$, $\mathbf{v} = -2\mathbf{i} + 5\mathbf{j}$ b. $\mathbf{u} = 3\mathbf{i} - 2\mathbf{j}$, $\mathbf{v} = 4\mathbf{i} + 6\mathbf{j}$

What is the definition of **The Alternate Form of the Dot Product of Two Vectors**?

Section 9.5 Objective 3 Using the Dot Product to Determine if Two Vector Are Orthogonal or Parallel

What is the definition of **Orthogonal Vectors**?

What is the **Test for Orthogonal Vectors**?

391

Section 9.5

Work through Example 4 and show all work below.

Determine the value of b so that the vectors $\mathbf{u} = \mathbf{i} + b\mathbf{j}$ and $\mathbf{v} = 3\mathbf{i} + 10\mathbf{j}$ are orthogonal.

What is the **Test for Parallel Vectors**?

Work through the video with Example 5 and show all work below.

Determine if $\mathbf{u} = -\dfrac{1}{2}\mathbf{i} - \mathbf{j}$ and $\mathbf{v} = 2\mathbf{i} + 4\mathbf{j}$ are orthogonal, parallel, or neither.

Section 9.5 Objective 4 Decomposing a Vector into Two Orthogonal Vectors

Describe the idea of **vector projections.**

What is the definition of **The Scalar Component of v in the Direction of w**?

392

What is the definition of **The Vector Projection of v onto w**?

Work through Example 6 and show all work below.

If $\mathbf{v} = \mathbf{i} + 2\mathbf{j}$ and $\mathbf{w} = 4\mathbf{i} + \mathbf{j}$, determine the vector $\text{proj}_{\mathbf{w}} \mathbf{v}$.

What is the definition of **The Vector Decomposition of v into Orthogonal Components**?

Work through the video with Example 7 and show all work below.

Let $\mathbf{v} = \mathbf{i} + 2\mathbf{j}$ and $\mathbf{w} = 4\mathbf{i} + \mathbf{j}$. Determine the vector decomposition of $\mathbf{v}$ into orthogonal components $\mathbf{v}_1$ and $\mathbf{v}_2$, where $\mathbf{v}_1$ is parallel to $\mathbf{w}$ and $\mathbf{v}_2$ is orthogonal to $\mathbf{w}$.

Section 9.5

Section 9.5 Objective 5 Solving Applications Involving Forces on an Inclined Plane

Work through the video with Example 8 and show all work below.
    A 200-pound object is placed on a ramp that is inclined at $22°$. What is the
    magnitude of the force needed to hold the box in a stationary position to prevent the
    box from sliding down the ramp? What is the magnitude of the force pushing against
    the ramp?

Section 9.5 Objective 6 Solving Applications Involving Work

What is the definition of **work**?

Work through Example 9 and show all work below.
    A horse is pulling a plow with a force of 400 pounds. The angle between the harness
    and the ground is $20°$. How much work is done to pull the plow 50 feet?

394

Section 10.1 Guided Notebook

Section 10.1 The Parabola

- ☐ Work through Objective 1
- ☐ Work through Objective 2
- ☐ Work through Objective 3
- ☐ Work through Objective 4
- ☐ Work through Objective 5

Introduction to Conic Sections

How are conic sections formed? Draw a picture below.

What are the four conic sections that can be formed?

View the animations for the circle and take notes below.

Section 10.1

View the animation for the ellipse and take notes below.

View the animation for the hyperbola and take notes below.

What are degenerate conic sections?

What are the three degenerate conic sections?

Section 10.1 The Parabola

Work through TTK #1 **finding the distance between Two Points Using the Distance Formula** and take notes here.

Work through TTK #2 **Converting the General Form of a Circle into Standard Form** and take notes here.

Section 10.1

Work through TTK #3 **Finding the Equations of Horizontal and Vertical Lines** and take notes here.

Section 10.1 Objective 1: Determining the Equation of a Parabola with a Vertical Axis of Symmetry

View the animation and describe the characteristics of a parabola.

What is the **geometric definition of a parabola**?

What is the equation of a parabola in standard form with a vertical axis of symmetry?

What is the vertex?

What is the distance from the vertex to the focus?

What is the distance from the vertex to the directrix?

What is the focus?

What is the equation of the directrix?

Draw two sketches of a vertical parabola and label each of the above.

Section 10.1

Work through Example 1 taking notes below.

Find the vertex, focus and directrix of the parabola $x^2 = 8y$ and sketch its graph.

Work through the video for Example 2 and take notes below.

Find the vertex, focus, and directrix of the parabola $-(x+1)^2 = 4(y-3)$ and sketch its graph.

Section 10.1 Objective 2: Determining the Equation of a Parabola with a Horizontal Axis of Symmetry.

What is the equation of a parabola in standard form with a horizontal axis of symmetry?

What is the vertex?

What is the distance from the vertex to the focus?

What is the distance from the vertex to the directrix?

What is the focus?

What is the equation of the directrix?

Draw two sketches of a horizontal parabola and label each of the parts.

Section 10.1

Work through the video with Example 3 and take notes below.

> Find the vertex, focus, and directrix of the parabola $(y - 3)^2 = 8(x + 2)$ and sketch its graph.

Section 10.1 Objective 3: Determining the Equation of a Parabola Given Information about the Graph.

Work through the video with Example 4 and take notes below.

> Find the standard form of the equation of the parabola with focus $\left(-3, \dfrac{5}{2}\right)$ and directrix $y = \dfrac{11}{2}$.

Work through the video with Example 5 and take notes below.

Find the standard form of the equation of the parabola with focus $(4,-2)$ and vertex $(\frac{13}{2},-2)$.

Section 10.1 Objective 4: Completing the Square to Find the Equation of a Parabola in Standard Form

Work through the video with Example 6 and take notes below.

Find the vertex, focus, and directrix and sketch the graph of the parabola $x^2-8x+12y=-52$.

Section 10.1

Section 10.1 Objective 5: Solving Applied Problems Involving Parabolas

What did the Romans create using properties of parabolic structures?

What are some of the objects that are manufactured that use parabolic surfaces?

Work through Example 7 and take notes below.

>   Parabolic microphones can be seen on the sidelines of professional sporting events so
>   that television networks can capture audio sounds from the players on the field. If the
>   surface of a parabolic microphone is 27 centimeters deep and has a diameter of 72
>   centimeters at the top, where should the microphone be placed relative to the vertex
>   of the parabola?

Section 10.2 Guided Notebook

Section 10.2 The Ellipse

- ☐ Work through Objective 1
- ☐ Work through Objective 2
- ☐ Work through Objective 3
- ☐ Work through Objective 4

Section 10.2 The Ellipse

View the animation for the ellipse and write the geometric definition of the ellipse. Draw a sketch below demonstrating the definition.

405

Section 10.2

Section 10.2 Objective 1: Sketching the Graph of an Ellipse

Work through the video describing horizontal and vertical ellipses. Draw a sketch and label the key features of each below.

What is the equation of an ellipse in standard form with center (h, k) and a horizontal major axis?

What is the relationship between a and b?

What are the ordered pairs of the foci?

What are the ordered pairs of the vertices?

What are the endpoints of the major axis?

What is the relationship between a, b, and c?

Draw a sketch labeling all from the previous page.

What is the equation of an ellipse in standard form with the center at the origin with a **horizontal** major axis? What is the equation of an ellipse in standard form with the center at the origin with a **vertical** major axis? Label the vertices, endpoints, and foci.

Work through the video with Example 1 and take notes below.

Sketch the graph of the ellipse $\dfrac{x^2}{25} + \dfrac{y^2}{4} = 1$, and label the center, foci, and vertices.

Work through the video with Example 2 and take notes below.

Sketch the graph of the ellipse $\dfrac{(x+2)^2}{20} + \dfrac{(y-3)^2}{36} = 1$, and label the center, foci, and vertices.

Section 10.2 Objective 2: Determining the Equation of an Ellipse Given Information about the Graph

Work through the video with Example 3 and take notes below.

> Find the standard form of the equation of the ellipse with foci at (-6,1) and (-2,1) such that the length of the major axis is eight units.

Work through the video with Example 4 and take notes below.

> Determine the equation of the ellipse with foci located at (0,6) and (0,-6) that passes through the point (-5,6).

Section 10.2

Section 10.2 Objective 3: Completing the Square to Find the Equation of an Ellipse in Standard Form

Work through the video with Example 5 and take notes below.

Find the center and foci and sketch the ellipse

$$36x^2 + 20y^2 + 144x - 120y - 396 = 0.$$

Section 10.2 Objective 4: Solving Applied Problems Involving Ellipses

What is the first application mentioned in the text as an application of ellipses?

What is sound wave lithotripsy?

Work through Example 6 and take notes below.

A patient is placed in an elliptical tank that is 200 centimeters long and 80 centimeters wide to undergo sound wave lithotripsy treatment for kidney stones. Determine where the sound emitter and the stone should be positioned relative to the center of the ellipse.

Section 10.3 The Hyperbola

- ☐ Work through TTK #1
- ☐ Work through TTK #2
- ☐ Work through TTK #3
- ☐ Work through Objective 1
- ☐ Work through Objective 2
- ☐ Work through Objective 3
- ☐ Work through Objective 4

Section 10.3 The Hyperbola

View the animation for a hyperbola and take notes below.

What is the geometric definition of a hyperbola?

Section 10.3 Objective 1: Sketching the Graph of a Hyperbola

Work through the video accompanying Objective 1 and describe each of the following:

Center:

Vertices:

Foci:

Transverse axis:

Conjugate Axis:

Asymptotes:

Reference rectangle:

Draw a sketch of two different hyperbolas below labeling these items.

413

Section 10.3

Write down **Fact 1 for Hyperbolas**.

What is the equation of a Hyperbola in Standard form with a Horizontal Transverse Axis?

      What are the ordered pairs of the foci?

      What are the ordered pairs of the vertices?

      What are the ordered pairs of the endpoints?

      What is the relationship between a, b and c?

      What is the equation of the asymptotes?

What is the equation of a Hyperbola in Standard form with a Vertical Transverse Axis?

      What are the ordered pairs of the foci?

      What are the ordered pairs of the vertices?

      What are the ordered pairs of the endpoints?

      What is the relationship between a, b and c?

      What is the equation of the asymptotes?

What is the equation of a Hyperbola in Standard from with a Horizontal Transverse Axis and center at the origin?

What are the ordered pairs of the foci?

What are the ordered pairs of the vertices

What are the ordered pairs of the endpoints?

What is the relationship between a, b and c ?

What is the equation of the asymptotes?

What is the equation of a Hyperbola in Standard from with a Vertical Transverse Axis and center at the origin?

What are the ordered pairs of the foci?

What are the ordered pairs of the vertices?

What are the ordered pairs of the endpoints?

What is the relationship between a, b and c?

What is the equation of the asymptotes?

Work through the video with Example 1 and take notes below.

Sketch the following hyperbolas. Determine the center, transverse axis, vertices, and foci and find the equations of the asymptotes.

a. $\dfrac{(y-4)^2}{36} - \dfrac{(x+5)^2}{9} = 1$

b. $25x^2 - 16y^2 = 400$

Section 10.3 Objective 2: Determining the Equation of a Hyperbola in Standard Form

Work through the video with Example 2 and take notes below.

Find the equation of the hyperbola with the center at (-1,0), a focus at (-11,0), and a vertex at (5,0).

Section 10.3

<u>Section 10.3 Objective 3: Completing the Square to Find the Equation of a Hyperbola in Standard Form</u>

Work through the video with Example 3 and take notes below.

Find the center, vertices, foci, and equations of asymptotes and sketch the hyperbola $12x^2 - 4y^2 - 72x - 16y + 140 = 0$.

Section 10.3 Objective 4: Solving Applied Problems Involving Hyperbolas

List 3 examples of applications of hyperbolas.

Work through Example 4 and take notes below.

One transmitting station is located 100 miles due east from another transmitting station. Each station simultaneously sends out a radio signal. The signal from the west tower is received by a ship $\dfrac{1,600}{3}$ microseconds after the signal from the east tower. If the radio signal travels at 0.18 miles per microsecond, find the equation of the hyperbola on which the ship is presently located.

Section 11.1 Guided Notebook

Section 11.1 Systems of Linear Equations in Two Variables

- ☐ Work through TTK #1
- ☐ Work through TTK #2
- ☐ Work through TTK #3
- ☐ Work through TTK #4
- ☐ Work through Objective 1
- ☐ Work through Objective 2
- ☐ Work through Objective 3
- ☐ Work through Objective 4

Section 11.1 Systems of Linear Equations in Two Variables

Work through TTK #1 **Applications of Linear Equations** and take notes below.

Work through TTK #2 **Writing the Equation of a Line in Standard Form** and take notes below.

Work through TTK #3 **Sketching Lines by Plotting Intercepts** and take notes below.

Work through TTK #4 **Understanding the Definition of Parallel Lines** and take notes below.

Section 11.1

Section 11.1 Objective 1: Verifying Solutions to a System of Linear Equations in Two Variables

Write down the definition of a linear equation in n variables.

What is the important thing to remember about the variables of a linear equation?

Work through Example 1 and take notes below.

Show that the ordered pair (-1,3) is a solution to the system

$3x - 2y = -9$
$x + y = 2$.

What is a consistent system? Draw a picture of a consistent system.

What is an inconsistent system? Draw a picture of an inconsistent system.

What are independent equations? Draw two pictures of independent equations.

Section 11.1

What are dependent equations? Draw a picture of dependent equations.

Section 11.1 Objective 2: Solving a System of Linear Equations Using the Substitution
Method

Watch the video on substitution and take notes below.

.

What are the steps for solving a system of equations by the method of substitution?

424

Work through Example 2 and take notes below.

Solve the following system using the method of substitution:

$$2x - 3y = -5$$
$$x + y = 5$$

Section 11.1 Objective 3: Solving a System of Linear Equations Using the Elimination
Method

Watch the video on the elimination method and take notes below.

What are the steps for solving a system of equations by the method of elimination?

Section 11.1

Work through Example 3 and take notes below.

Solve the following system using the method of elimination:

$-2x + 5y = 29$
$3x + 2y = 4$.

Work through Example 4 and take notes below.

Solve the system $\begin{array}{l} x - 2y = 11 \\ -2x + 4y = 8 \end{array}$.

What indicates that the problem in Example 4 is an inconsistent system?

Why, geometrically, does the system have no solution?

Work through Example 5 and take notes below.

Solve the system $\begin{array}{l} -3x+6y=9 \\ x-2y=-3 \end{array}$.

What is the solution of this system geometrically?

What are the two ways the solution can be expressed using ordered pairs?

Section 11.1

Section 11.1 Objective 4: Solving Applied Problems Using a System of Linear Equations

What is the five step strategy for problem solving using systems of equations?

Work through Example 6 and take notes below. Be sure to use the two variable system.

Roger Staubach and Terry Bradshaw were both quarterbacks in the National Football League. In 1973, Staubach threw three touchdown passes more than twice the number of touchdown passes thrown by Bradshaw. If the total number of touchdown passes between Staubach and Bradshaw was 33, how many touchdown passes did each player throw?

Work through Example 7 and take notes below.

> During one night at the jazz festival, 2,100 tickets were sold. Adult tickets sold for
> $12, and child tickets sold for $7. If the receipts totaled $22,100, how many of each
> type of ticket were sold?

Work through the video with Example 8 and take notes below.

> Twin City Foods, Inc., created a 10-lb bean mixture that sells for $5.75 by mixing
> lima beans and green beans. If lima beans sell for $.70 per pound and green beans
> sell for $.50 per pound, how many pounds of each bean went into the mixture?

Section 11.1

Work through the video with Example 9 and take notes below.

A small airplane flies from Seattle, Washington, to Portland, Oregon – a distance of 150 miles. Because the pilot encountered a strong headwind, the trip took 1 hour and 15 minutes. On the return flight, the wind is still blowing at the same speed. If the return trip took 45 minutes, what was the average speed of the airplane in still air? What was the speed of the wind?

What is the formula that relates distance, time and rate?

What affect does the headwind have on the speed of the plane?

What affect does the tailwind have on the speed of the plane?

Why does the time need to be in hours?

Show the complete solution to the problem.

Section 11.2 Guided Notebook

Section 11.2 Systems of Linear Equations in Three Variables

☐ Work through TTK #1
☐ Work through TTK #2
☐ Work through Objective 1
☐ Work through Objective 2
☐ Work through Objective 3
☐ Work through Objective 4

Section 11.2 Systems of Linear Equations in Three Variables

Work through TTK #1 **Solving a System of Linear Equations Using the Substitution Method** and take notes below.

Work through TTK #2 **Solving a System of Linear Equations Using the Elimination Method** and take notes below.

Section 11.2

What is the solution to a system of three linear equations in three variables if it is unique?

Work through the video with Example 1 and take notes below.

Verify that the ordered triple (1, 2, 1) is a solution to the following system of linear equations:

$$2x + 3y + 4z = 12$$
$$x - 2y + 3z = 0$$
$$-x + y - 2z = -1$$

Section 11.2 Objective 2: Solving a System of Linear Equations Using Gaussian Elimination

What is triangular form?

What is back substitution?

Who was Gaussian elimination named after?

Write down the three elementary row operations.

What does R_i describe?

Write down the notation used to describe elementary row operations.

Section 11.2

Work through the video with Example 2 taking notes below.

For the following system, use elementary row operations to find an equivalent system in triangular form and then use back substitution to solve the system:

$$2x + 3y + 4z = 12$$
$$x - 2y + 3z = 0$$
$$-x + y - 2z = -1$$

Section 11.2 Objective 3: Using an Augmented Matrix to Solve a System of Linear Equations

What is an augmented matrix?

Work through the video with Example 3 and take notes below.

> Create an augmented matrix and solve the following linear system using Gaussian elimination by writing an equivalent system in triangular form:
>
> $$x + 2y - z = 3$$
> $$x - 3y - 2z = 11$$
> $$-x - 2y + 2z = -6$$

Work through the video on triangular, row-echelon and reduced row-echelon form. Take notes and answer the questions below.

What is triangular form?

What is row-echelon form?

What is reduced row-echelon form?

What is the benefit of reduced row-echelon form?

What is Gauss-Jordan elimination?

Work through the video with Example 4 and take notes below.

Solve the following system using Gauss-Jordan elimination:

$$x_1 + x_2 + x_3 = -1$$
$$x_1 + 2x_2 + 4x_3 = 3$$
$$x_1 + 3x_2 + 9x_3 = 3$$

Section 11.2 Objective 4: Applications of Systems of Linear Equations Involving Three Variables

Work through the video with Example 5 and take notes below.

Wendy ordered 30 T-shirts online for her three children. The small T-shirts costs $4 each, the medium T-shirts costs $5 each, and the large T-shirts were $6 each. She spent $40 more purchasing the large T-shirts than the small T-shirts. Wendy's total bill was $154. How many T-shirts of each size did buy?

Section 11.2

How many points determine the equation of a line?

How many non-collinear points determine the equation of a quadratic function?

Work through Example 6 take notes below.

Determine the quadratic function whose graph passes through the three points (1,-9), (-1,-5), and (-3,7).

Section 11.3 Guided Notebook

Section 11.3 Inconsistent and Dependent Linear Systems in Three Variables

- ☐ Work through TTK #1
- ☐ Work through Objective 1
- ☐ Work through Objective 2
- ☐ Work through Objective 3

Section 11.3 Inconsistent and Dependent Linear Systems in Three Variables

Work through TTK #1 **Solving a System of Three Equations in Three Variables Using Gauss-Jordan Elimination** and show work below.

What is a **consistent system**?

What is an **inconsistent system**?

Section 11.3

Section 11.3 Objective 1: Solving Linear Systems Having No Solutions

Work through the video with Example 1 and take notes below.

Use Gauss-Jordan elimination to solve the system:

$$x - y + 2z = 4$$
$$-x + 3y + z = -6$$
$$x + y + 5z = 3$$

How can you tell this system is inconsistent?

How many solutions does this system have?

Section 11.3 Objective 2: Solving Linear Systems Having Infinitely Many Solutions

Work through the video with Example 2 and take notes below.

Use Gauss-Jordan elimination to solve the system:

$$x + 2y + 3z = 10$$
$$x + y + z = 7$$
$$3x + 2y + z = 18$$

Why is this a dependent system?

How many solutions does this system have?

Section 11.3

Work through the video with Example 3 and take notes below.

Each augmented matrix in row reduced form in equivalent to the augmented matrix of a system of linear equations in variables, x, y, and z. Determine whether the system is dependent or inconsistent. If the system is dependent describe the solution as in Example 2.

a. $\begin{bmatrix} 1 & 0 & -2 & | & 5 \\ 0 & 1 & 3 & | & -2 \\ 0 & 0 & 0 & | & 0 \end{bmatrix}$
b. $\begin{bmatrix} 1 & 0 & 0 & | & -4 \\ 0 & 1 & 0 & | & 6 \\ 0 & 0 & 0 & | & 10 \end{bmatrix}$
c. $\begin{bmatrix} 1 & 0 & 0 & | & 3 \\ 0 & 1 & -2 & | & 4 \\ 0 & 0 & 0 & | & 0 \end{bmatrix}$

Section 11.3 Objective 3: Solving Linear Systems Having Fewer Equations Than Variables

What are the geometric possibilities for a system of three variables but only two equations?

Work through Example 4 taking notes below.

Solve the linear system using Gauss-Jordan elimination:

$$x + y + z = 1$$

$$2x - 2y + 6z = 10$$

Section 11.4 Guided Notebook

Section 11.4 Partial Fraction Decomposition

- ☐ Work through TTK #1
- ☐ Work through TTK #2
- ☐ Work through TTK #3
- ☐ Work through Objective 1
- ☐ Work through Objective 2
- ☐ Work through Objective 3
- ☐ Work through Objective 4

Section 11.4 Partial Fraction Decomposition

Work through TTK #1 **Solving a System of Linear Equations Using the Substitution Method** and show work below.

Work through TTK #2 **Solving a System of Linear Equations Using the Elimination Method** and show work below.

Work through TTK #3 **Solving a System of Linear Equations Using Gaussian Elimination** and show work below.

What is **partial fraction decomposition**?

What are **partial fractions**?

What two criteria must be satisfied in order to perform partial fraction decomposition?

Section 11.4

<u>Section 11.4 Objective 1: Decomposing Rational Expressions of the Form $P(x)/Q(x)$, where $Q(x)$ has Only Distinct Linear Factors</u>

If the denominator of $P(x)/Q(x)$ is the product of n distinct linear factors, then the partial fraction decomposition of $P(x)/Q(x)$ will be of what form?

Work through Example 1 and take notes below.

Determine the partial fraction decomposition of

$$\frac{x+10}{2x^2+5x-3}.$$

In the short-cut method described in the note on page 11.4-8 of the eText, how do we choose values of x?

Using this method, solve for A, B, C in Example 1.

446

What are the seven **Steps for Determining the Partial Fraction Decomposition of P(x)/Q(x)?**

Work through the video with Example 2 and take notes below.

Determine the partial fraction decomposition of $\dfrac{5x^2 - x + 1}{x^3 + 3x^2 - 4x}$.

Section 11.4

Section 11.4 Objective 2: Decomposing Rational Expressions of the Form $P(x)/Q(x)$, where $Q(x)$ has a Repeated Linear Factor

If the denominator of $P(x)/Q(x)$ has a repeated factor, then for every repeated linear factor of the form $(ax+b)^n$, we introduce n partial fractions of what form?

Work through Example 3 and take notes below.

Set up the partial fraction decomposition for

$$\frac{x}{x^2(3x-1)^3(x-5)}.$$ Do not solve for the constants.

Work through the video with Example 4 and take notes below.

Determine the partial fraction decomposition of $\dfrac{x-1}{x(x-2)^2}$.

Section 11.4 Objective 3: Decomposing Rational Expressions of the Form *P(x)/Q(x)*, where *Q(x)* has a Distinct Prime Quadratic Factor

If the denominator of *P(x)/Q(x)* has a prime quadratic factor, then for every prime quadratic factor of the form $ax^2 + bx + c$, we introduce n partial fractions of what form?

Work through Example 5 and take notes below.

Set up the partial fraction decomposition for $\dfrac{5x^3 - 7x^2 - 8x + 1}{(x^2 + 1)(2x^2 + x + 7)(5x - 3)^2}$. Do not solve for the constants.

Work through the video with Example 6 and take notes below.

Determine the partial fraction decomposition of $\dfrac{8x^2 + 7}{x^3 - 1}$.

Section 11.4

Section 11.4 Objective 4: Decomposing Rational Expressions of the Form $P(x)/Q(x)$, where $Q(x)$ has a Repeated Prime Quadratic Factor

If the denominator of $P(x)/Q(x)$ has a repeated prime quadratic factor, then for every repeated prime quadratic factor of the form $(ax^2 + bx + c)^n$, we introduce n partial fractions of what form?

Work through Example 7 taking notes below.

Set up the partial fraction decomposition for $\dfrac{x^4 + x^3 + x^2 + x + 1}{(x^2 + 4)^3 (3x^2 + 10x + 1)^2}$. Do not solve for the constants.

Work through the video with Example 8 and take notes below.

Determine the partial fraction decomposition of $\dfrac{3x^4 - 5x^3 + 7x^2 + x - 2}{(x-1)(x^2+1)^2}$.

Section 11.5 Guided Notebook

Section 11.5 Systems of Nonlinear Equations

- ☐ Work through Objective 1
- ☐ Work through Objective 2
- ☐ Work through Objective 3
- ☐ Work through Objective 4

Section 11.5 Systems of Nonlinear Equations

What is a **nonlinear system**?

How can we graphically represent the real solutions to a nonlinear system?

Section 11.5 Objective 1: Determining the Number of Solutions to a System of Nonlinear Equations

Work through the interactive video with Example 1 and take notes below.

For each system of nonlinear equations, sketch the graph of each equation of the system and then determine the number of real solutions to each system. Do not solve the system.

a. $\begin{aligned} x^2 + y^2 &= 25 \\ x - y &= 1 \end{aligned}$ **b.** $\begin{aligned} x - y^2 &= 4 \\ x - y &= 6 \end{aligned}$ **c.** $\begin{aligned} x^2 + y^2 &= 9 \\ x^2 - y &= 3 \end{aligned}$

Section 11.5 Objective 2: Solving a System of Nonlinear Equations Using the Substitution Method

What are the five steps for **Solving a System of Nonlinear Equations by the Substitution Method**?

Work through the video with Example 2 and take notes below.

Determine the real solutions to the following system using the substitution method.

$$x^2 + y^2 = 25$$
$$x - y = 1$$

After obtaining our proposed solutions to a system of nonlinear equations, what is it absolutely critical to do?

Section 11.5

Work through the video with Example 3 and take notes below.

Determine the real solutions to the following system using the substitution method.

$$5x^2 - y^2 = 25$$
$$2x + y = 0$$

Work through the video with Example 4 and take notes below.

Determine the real solutions to the following system using the substitution method.

$$x^2 + 2y^2 = 18$$
$$xy = 4$$

Section 11.5 Objective 3: Solving a System of Nonlinear Equations Using Substitution, Elimination, or Graphing

What are the six steps for **Solving a System of Nonlinear Equations by the Elimination Method**?

Work through the interactive video with Example 5 and take notes below.

Determine the real solutions to the following system.

$$x^2 + y^2 = 9$$
$$x^2 - y = 3$$

Section 11.5

Work through the video with Example 6 and take notes below.

Determine the real solutions to the following system.

$$y = \log(3x+1) - 5$$
$$y = \log(x-2) - 4$$

Work through Example 7 and take notes below.

Determine the real solutions to the following system.

$$y = \ln x$$
$$y = 1 - x$$

<u>Section 11.5 Objective 4: Solving Applied Problems Using a System of Nonlinear Equations</u>

Work through Example 8 taking notes below.

Find two positive numbers such that the sum of the squares of the two numbers is 25 and the difference between the two numbers is 1.

Work through Example 9 taking notes below.

An open box (with no lid) has a rectangular base. The height of the box is equal in length to the shortest side of the base. What are the dimensions of the box if the volume is 176 cubic inches and the surface area is 164 square inches?

Section 11.6 Guided Notebook

Section 11.6 Systems of Inequalities

- ☐ Work through Objective 1
- ☐ Work through Objective 2
- ☐ Work through Objective 3
- ☐ Work through Objective 4
- ☐ Work through Objective 5
- ☐ Work through Objective 6

Section 11.6 Systems of Inequalities

<u>Section 11.6 Objective 1: Determine If an Ordered Pair is a Solution to an Inequality in Two Variables</u>

Given an inequality in the variables x and y, we can determine if the ordered pair (a, b) is a solution to the inequality by doing what?

Work through the video with Example 1 and take notes below.

Determine if the given ordered pair is a solution to the inequality $-2x + y^2 \le 4$.

a. $(2, -1)$ **b.** $(-2, 0)$ **c.** $(-\frac{1}{2}, \frac{7}{2})$

Section 11.6 Objective 2: Graphing a Linear Inequality in Two Variables

What is the definition of a **linear inequality in two variables**?

What is a **boundary line**?

What are the three **Steps for Graphing Linear Inequalities in Two Variables**?

Work through the interactive video with Example 2 and take notes below.

Graph each inequality.

    a. $x - 2y \geq 4$ **b.** $3y < 2x$ **c.** $x < -2$

Section 11.6

What are the three **Steps for Graphing Nonlinear Inequalities in Two Variables**?

According to the note on page 11.6-14, in order to be certain that all of the solution set has been identified it may be necessary to do what? When is this necessary?

Work through the interactive video with Example 3 and take notes below.

Graph each inequality.

    a. $x^2 + y^2 \geq 9$ **b.** $9y^2 - 4x^2 < 36$ **c.** $x - y^2 \leq 1$

Section 11.6 Objective 4: Determining if an Ordered Pair is a Solution to a System of Inequalities in Two Variables

What is the definition of a **system of inequalities in two variables**?

Work through the video with Example 4 taking notes below.

Determine which ordered pairs are solutions to the given system.

a. $\begin{aligned} 2x - 3y \le 9 \\ 2x - y \ge -1 \end{aligned}$ i. (1,-2)   ii. (-1,2)   iii. (3,-1)

b. $\begin{aligned} x^2 - y^2 \le 25 \\ x - y > 1 \end{aligned}$ i. (-1,-3)   ii. (0,5)   iii. (-3,4)

Section 11.6 Objective 5: Graphing a System of Linear Inequalities in Two Variables

Work through the animation with Objective 5 and take notes below.

Graph the following system of linear inequalities.

$$2x - 3y \le 9$$
$$2x - y \ge -1$$

461

Section 11.6

What are the two **Steps for Graphing Systems of Linear Inequalities in Two Variables?**

Work through the interactive video with Example 5 and take notes below.

Graph each system of linear inequalities in two variables.

a. $\begin{array}{l} x+y>2 \\ 2x-y\le 6 \end{array}$ **b.** $\begin{array}{l} x-3y>6 \\ 2x-6y<-9 \end{array}$ **c.** $\begin{array}{l} 4x>y \\ x-3y<-9 \\ x+y<4 \end{array}$

Section 11.6 Objective 6: Graphing a System of Nonlinear Inequalities in Two Variables

What are the two **Steps for Graphing Systems of Nonlinear Inequalities in Two Variables**?

Work through the interactive video with Example 6 and take notes below.

Graph each system of nonlinear inequalities in two variables.

a. $\begin{array}{l} x^2 + y^2 \le 25 \\ x - y > 1 \end{array}$ **b.** $\begin{array}{l} x - y^2 \ge 4 \\ x - y \le 6 \end{array}$ **c.** $\begin{array}{l} 4x^2 + 9y^2 \le 36 \\ x^2 + y \le -2 \end{array}$

Section 12.1 Guided Notebook

Section 12.1 Matrix Operations
Work through TTK #1
Work through Objective 1
Work through Objective 2
Work through Objective 3
Work through Objective 4
Work through Objective 5

Section 12.1 Matrix Operations

12.1 Things To Know

1. Solving a System of Three Equations in Three Variables Using Gauss-Jordan Eliminations.

Try working through a "You Try It" problem.

Section 12.1 Objective 1 Understanding the Definition of a Matrix

Watch the video with Objective 1, take notes below and answer the following questions.

What is the purpose of the vertical bar?

Write down the definition of a matrix.

When are two matrices considered to be equal?

What is typically used to name a matrix?

Write an example of each of the following:

Row matrix:

Column matrix:

Identity matrix:

3 x 3 matrix:

2 x 2 matrix:

3 x 2 matrix:

What is a square matrix?

What is a row matrix?

What is a column matrix?

What is an identity matrix?

Section 12.1 Objective 2 Adding and Subtracting Matrices

In order to add or subtract two matrices they must be _____?

What is the procedure for adding or subtracting matrices?

Work through Example 1 showing all work below.

Let $A = \begin{bmatrix} 2 & -1 & 0 & 3 \\ -1 & 4 & -2 & 5 \\ 5 & -3 & 7 & 9 \end{bmatrix}$ and $B = \begin{bmatrix} -3 & -2 & 6 & 1 \\ 0 & 4 & -5 & 1 \\ -1 & 4 & 2 & 7 \end{bmatrix}$

    a. Find matrix $A + B$
    b. Find matrix $B + A$
    c. Find matrix $A - B$

Section 12.1

Section 12.1 Objective 3 Scalar Multiplication of Matrices

What is a scalar?

Work through the video with Example 2 and take notes below.

If $A = \begin{bmatrix} -4 & 0 \\ 3 & 1 \end{bmatrix}$ and $B = \begin{bmatrix} 6 & 3 \\ -3 & -9 \end{bmatrix}$, find the following matrices; $-2A$, $\frac{1}{3}B$, and $-2A +$

$\frac{1}{3}B$.

Section 12.1 Objective 4 Multiplication of Matrices.

Can any two matrices be multiplied?

If not, when can two matrices be multiplied?

Write down the procedure for Matrix Multiplication.

Watch the video with Example 3 and show all steps below.

Given matrices $A = \begin{bmatrix} 1 & 1 & 2 \\ 2 & 4 & -3 \\ 3 & 6 & -5 \end{bmatrix}, B = \begin{bmatrix} 0 & -1 & 1 \\ -2 & 1 & 3 \\ 4 & 5 & -3 \end{bmatrix}, and\ C = \begin{bmatrix} 1 & 0 & 0 \\ 0 & 1 & 0 \\ 0 & 0 & 1 \end{bmatrix}$, find the

products AB, BA and AC.

Section 12.1

Is matrix multiplication commutative?

What is the notation for a square identity matrix?

What is the Identity Property for Matrix Multiplication?

Work through Example 4 taking notes below.

If $A = \begin{bmatrix} 2 & -1 \\ 5 & 6 \\ 0 & -3 \end{bmatrix}$, find appropriate identity matrices I_m and I_n and verify that $I_m A = A$

and $AI_n = A$

Write down the Summary of Matrix Operations.

Section 12.1 Objective 5 Applications of Matrix Multiplication

Describe the data in the matrix with Objective 5.

Section 12.2 Guided Notebook

Section 12.2 Inverses of Matrices and Matrix Equations
    Work through TTK #1
    Work through TTK #2
    Work through TTK #3
    Work through TTK #4
    Work through Objective 1
    Work through Objective 2
    Work through Objective 3
    Work through Objective 4

Section 12.2 Inverses of Matrices and Matrix Equations

12.2 Things to Know

1. Solving a System of Three Equations in Three Variables Using Gauss-Jordan Elimination.

Try working through a "You Try It" problem.

2. **Scalar Multiplication of Matrices**. Try working through a "You Try It" problem.

3. **Multiplication of Matrices**. Try working through a "You Try It" problem.

4. **Applications of Matrix Multiplication**. Try working through a "You Try It" problem.

Section 12.2

<u>Section 12.2 Objective 1 Understanding the Definition of an Inverse Matrix.</u>

What is the Multiplicative Inverse of a Matrix?

Work through the video with Example 1 showing all steps below.

Verify that $A = \begin{bmatrix} 1 & 1 & 2 \\ 2 & 4 & -3 \\ 3 & 6 & -5 \end{bmatrix}$ and $B = \begin{bmatrix} 2 & -17 & 11 \\ -1 & 11 & -7 \\ 0 & 3 & -2 \end{bmatrix}$ are inverse matrices.

Section 12.2 Objective 2 Finding the Inverse of a 2 x 2 Matrix Using a Formula.

For a matrix to have an inverse matrix it must be _____?

Does every square matrix have an inverse?

When is a matrix **invertible**?

When is a matrix **singular**?

What is a **determinant**?

What is the formula for the **Determinant of a 2 x 2 Matrix**?

How is the determinant notated?

Using the determinant, when is a matrix invertible?

Section 12.2

Write down the **Formula for Determining the Inverse of a 2 x 2 Matrix.**

Watch the video of the proof and write down the proof below.

Watch the video with Example 2 and write down all the steps below.

Determine whether the following matrices are invertible or singular. If the matrix is invertible, find its inverse and then verify by using matrix multiplication.

a. $A = \begin{bmatrix} 2 & 3 \\ -4 & -6 \end{bmatrix}$

b. $B = \begin{bmatrix} 2 & -1 \\ 4 & 3 \end{bmatrix}$

<u>Section 12.2 Objective 3 Finding the Inverse of an Invertible Square Matrix</u>

What are the **Steps for Finding the Multiplicative Inverse of an Invertible Square Matrix**?

Work through Example 3 with the video. Show all steps below.

If possible, determine the inverse of $A = \begin{bmatrix} 1 & 1 & 2 \\ 2 & 4 & -3 \\ 3 & 6 & -5 \end{bmatrix}$.

Section 12.2

Work through Example 4 and take notes as you show all steps below.

If possible, determine the inverse of $A = \begin{bmatrix} 5 & 0 & -1 \\ 1 & -3 & -2 \\ 0 & 5 & 3 \end{bmatrix}$.

What is the caution statement after Example 4?

Section 12.2 Objective 4 Solving Systems of Equations Using an Inverse Matrix

Work through the video with Objective 4 and take detailed notes below.

What is a coefficient matrix?

What is a variable matrix?

What is a constant matrix?

Is matrix multiplication commutative?

Because of this where must the inverse matrix be positioned?

Section 12.2

Work through the video with Example 5 showing all steps below.

Solve the following system of linear equations using an inverse matrix:

$$-3x + 2y = 4$$
$$5x - 4y = 9$$

Work through Example 6 and take notes below.

Solve the following system of linear equations using an inverse matrix:

$$2x + 4y + z = 3$$
$$-x + y - z = 6$$
$$x + 4y \quad = 7$$

Section 12.3 Determinants and Cramer's Rule
Work through TTK #1

Work through TTK #2

Work through TTK #3

Work through Objective 1

Work through Objective 2

Work through Objective 3

Section 12.3 Determinants and Cramer's Rule

12.3 Things To Know

1. **Solving a System of Three Equations in Three Variables Using Gauss-Jordan Elimination.**

Try working through a "You Try It" problem.

2. **Finding the Inverse of a 2 x 2 Matrix Using a Formula**. Try working through a "You Try It" problem.

3. **Solving Systems of Equations Using an Inverse Matrix**. Try working through a "You Try It" problem.

Section 12.3

Section 12.3 Objective 1 Using Cramer's Rule to Solve Linear Systems in Two Variables

Write down **Cramer's Rule for a System of Linear Equations in Two Variables**.

Write down the proof of Cramer's Rule.

Work through the video with Example 1 showing all steps below.
Use Cramer's rule to solve the following system:

$$2x - 2y = 4$$
$$3x + y = -3$$

Section 12.3 Objective 2 Calculating the Determinant of an n x n Matrix by Expansion of Minors

How do you determine the 2 x 2 matrix associated with each entry of a 3 x 3 matrix?

What are minors?

Using the 3 x 3 matrix in the text with Objective 2, what is the 2 x 2 matrix associated with the entry 2?

What is the 2 x 2 matrix associated with the entry 3?

What is the minor for the entry -7?

485

How do you find the **Determinant of a 3 x 3 matrix Using Expansion by Minors?**

Work through the video with Example 2 and show all steps below.

$$\text{Calculate } |A| \text{ if } A = \begin{bmatrix} 2 & -17 & 11 \\ -1 & 11 & -7 \\ 0 & 3 & -2 \end{bmatrix}.$$

Does the first row have to be used when calculating the determinant by expanding the minors?

If not using the first row, what must you be very careful to do?

Work through the video with Example 3 and show all steps below.

Find $|A|$ if $A = \begin{bmatrix} 2 & -17 & 11 \\ -1 & 11 & -7 \\ 0 & 3 & -2 \end{bmatrix}$ by expanding the minors down the first column.

Work through the video with Example 4 and show all steps below.

Calculate $|A|$ if $A = \begin{bmatrix} 8 & 0 & 1 & 10 \\ -1 & 0 & 0 & -5 \\ 5 & 2 & -2 & 4 \\ -9 & 0 & 0 & -12 \end{bmatrix}$

Section 12.3

<u>Section 12.3 Objective 3 Using Cramer's Rule to Solve Linear Systems in n Variables.</u>

What is Cramer's Rule for a System of Linear Equations in n Variables?

Work through Example 5 with the interactive video. Show all steps below.

     Use Cramer's rule to solve the system

$$x + y + z = 2$$
$$x - 2y + z = 5.$$
$$2x + y - z = -1$$

Section 13.1 Guided Notebook

13.1 Introduction to Sequences and Series
Work through TTK #1
Work through Objective 1
Work through Objective 2
Work through Objective 3
Work through Objective 4
Work through Objective 5
Work through Objective 6

Section 13.1 Introduction to Sequences and Series

13.1 Things To Know

1. Using the Vertical Line Test (Section 2.1)

Can you determine if the graph represents a function? Try working through a "You Try It" problem or refer to Section 2.1 or watch the video.

Section 13.1 Objective 1 Writing the Terms of a Sequence

What is the definition of a **finite sequence**?

Section 13.1

What is the definition of an **infinite sequence**?

What are the **terms** of the sequence?

What is the definition of **The Factorial of a Non-Negative Integer**?

By definition, we write the zero factorial as _____ and it is equal to _____.

Work through Example 1 showing all work below.
   Write the first four terms of each sequence whose nth term is given.
   a. $a_n = 2n - 1$

   b.   b. $b_n = n^2 - 1$

   c.   c. $c_n = \dfrac{3^n}{(n-1)!}$

   d.   d. $d_n = (-1)^n 2^{n-1}$

490

What is an **alternating sequence?**

Section 13.1 Objective 2 Writing the Terms of a Recursive Sequence

Work through the interactive video with Example 2 and show all work below.
Write the first four terms of each of the following recursive sequences.

a. $a_1 = -3$, $a_n = 5a_{n-1} - 1$ for $n \geq 2$

b.   b. $b_1 = 2$, $b_n = \dfrac{(-1)^{n-1}n}{b_{n-1}}$ $for\ n \geq 2$

Work through Example 3 and show all work below.
The Fibonacci sequence is defined recursively by $a_n = a_{n-1} + a_{n-2}$ where $a_1 = 1$ and $a_2 = 1$. Write the first eight terms of the Fibonacci sequence.

Section 13.1

Section 13.1 Objective 3 Writing the General Term for a Given Sequence

Work through Example 4 and take notes here on the process for determining a pattern in a given sequence.

      Write a formula for the nth term of each infinite sequence, then use this formula to find the 8th term of the sequence.

a. $\dfrac{1}{1}, \dfrac{1}{2}, \dfrac{1}{3}, \dfrac{1}{4}, \dfrac{1}{5}, \dots$

b. $-\dfrac{2}{1}, \dfrac{4}{2}, -\dfrac{8}{6}, \dfrac{16}{24}, -\dfrac{32}{120}, \dots$

Section 13.1 Objective 4 Computing Partial Sums of a Series

What is the definition of a **finite series**?

What is the definition of an **infinite series**?

The sum of the first n terms of a series is called the nth _____ **sum** of the series and is denoted as S_n.

Work through Example 5 and note how you determined the indicated partial sums below.
Given the general term of each sequence, find the indicated partial sum.

a. $a_n = \dfrac{1}{n}$, find S_3

b. $b_n = (-1)^n 2^{n-1}$, find S_5

Section 13.1 Objective 5 Determining the Sum of a Finite Series Written in Summation Notation

Summation Notation:

Index of summation:

Lower limit of summation:

Upper limit of summation:

Is it necessary for the lower limit of the summation to start at 1?

Section 13.1

Work through the interactive video with Example 6 and take notes here:
    Find the sum of each finite series.

a. $\displaystyle\sum_{i=1}^{5} i^2$

b. $\displaystyle\sum_{j=2}^{5} \frac{j-1}{j+1}$

c. $\displaystyle\sum_{k=0}^{6} \frac{1}{k!}$

Section 13.1 Objective 6 Writing a Series Using Summation Notation

Work through Example 7 and take notes below.

Rewrite each series using summation notation. Use 1 as the lower limit of summation.

a. $2 + 4 + 6 + 8 + 10 + 12$

b. $1 + 2 + 6 + 24 + 120 + 720 + \ldots + 3{,}628{,}800$

Section 13.2 Guided Notebook

13.2 Arithmetic Sequences and Series
Work through TTK #1
Work through TTK #2
Work through TTK #3
Work through Objective 1
Work through Objective 2
Work through Objective 3
Work through Objective 4

Section 13.2 Arithmetic Sequences and Series

13.2 Things To Know

1. Work TTK #1; **Solving a System of Linear Equations Using the Substitution Method** and show all work below.

2. Work TTK #2; **Solving a System of Linear Equations Using the Elimination Method** and show all work below.

3. Work TTK #3; **Determining the Sum of a Finite Series Written in Summation Notation** and show all work below.

Section 13.2 Objective 1 Determining if a Sequence is Arithmetic

What is an **Arithmetic Sequence**?

In an arithmetic sequence, the *d* refers to _____.

In an arithmetic sequence the **general term**, or *n*th term, has the form $a_n =$ _____.

Section 13.2

Work through the interactive video with Example 1 and show all work below.

For each of the following sequences, determine if it is arithmetic. If the sequence is arithmetic, find the common difference.

a. 1,4,7,10,13,...

b. $b_n = n^2 - n$

c. $a_n = -2n + 7$

d. $a_1 = 14, a_n = 3 + a_{n-1}$

Refer to Figure 2 on page 13.2–4.

Note that every Arithmetic Sequence is a <u>linear function</u> whose <u>domain is the natural numbers</u>.

What does it mean that the points are **collinear**?

_____

When the common difference of an arithmetic sequence is <u>positive</u>, the terms of the sequence_____ and the graph is represented by a set of ordered pairs that lies along a line with a positive slope.

When the common difference of an arithmetic sequence is <u>negative</u>, the terms of the sequence_____ and the graph is represented by a set of ordered pairs that lies along a line with a negative slope.

498

Section 13.2 Objective 2 Finding the General Term or a Specific Term of an Arithmetic
Sequence

What is the **formula for the general term of an arithmetic sequence**?

Work through the interactive video with Example 2 and show all work below.

Find the general term of each arithmetic sequence, then find the indicated term of the
sequence.

a. $11,17,23,29,35,\ldots$; a_{50}

b. $2,0,-2,-4,-6,\ldots$; a_{90}

c. Find a_{31}

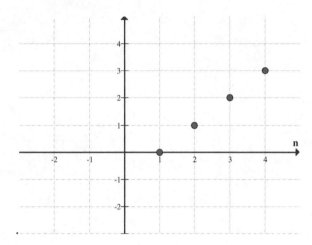

Section 13.2

Work through the interactive video with Example 3 and show all work below.
Make special note of part b where you are NOT given the common difference.
   a.   Given an arithmetic sequence with $d = -4$ and $a_3 = 14$, find a_{50}.
   b.   Given an arithmetic sequence with $a_4 = 12$ and $a_{15} = -10$, find a_{41}.

Section 13.2 Objective 3 Computing the nth Partial Sum of an Arithmetic Sequence

Make note of the formula for the ***nth Partial Sum of an Arithmetic Sequence***:

Work through Example 4 and show all work below.

Find the sum of each arithmetic series.

a. $\displaystyle\sum_{i=1}^{20}(2i-11)$ 　　　　b. $-5+(-1)+3+7+\ldots+39$

Section 13.2 Objective 4 Applications of Arithmetic Sequences and Series

Work through Example 5 and show all work below.

A local newspaper has hired teenagers to go door-to-door to try to solicit new subscribers. The teenagers receive \$2 for selling the first subscription. For each additional subscription sold, the newspaper will pay the teenagers 10 cents more than what was paid for the previous subscription. How much will the teenagers get paid for selling the 100th subscription? How much money will the teenagers earn by selling 100 subscriptions?

Work through the video with Example 6 and show all work below.

A large multiplex movie house has many theaters. The smallest theater has only 12 rows. There are six seats in the first row. Each row has two seats more than the previous row. How many total seats are there in this theater?

Section 13.3 Guided Notebook

13.3   Geometric Sequences and Series
Work through TTK # 1
Work through TTK # 2
Work through TTK # 3
Work through Objective 1
Work through Objective 2
Work through Objective 3
Work through Objective 4
Work through Objective 5
Work through Objective 6

Section 13.3 Geometric Sequences and Series

<u>**13.3 Things To Know**</u>

1. Work through TTK #1 **Using the Periodic Compound Interest Formula** and show all work below.

2. Work through TTK #2 **Finding the General Term or Specific Term of an Arithmetic Sequence** and show all work below.

3. Work through TTK #3 **Computing the *n*th Partial Sum of an Arithmetic Series** and show all work below.

Section 13.3 Objective 1 Writing the Terms of a Geometric Sequence

Write the definition of a **Geometric Sequence:**

How do you determine the **common ratio**?

State the formula for the ***n*th term** or the **general term of a geometric sequence**:

Work through the interactive video with Example 1 and show all work below.
  a. Write the first five terms of the geometric sequence having a first term of 2 and a common ratio of 3.
  b. Write the first five terms of the geometric sequence such that $a_1 = -4$ and $a_n = -5a_{n-1}$ for $n \geq 2$.

Section 13.3 Objective 2 Determining if a Sequence is Geometric

Work through the interactive video for Example 2 and show all work below.
  For each of the following sequences, determine if it is geometric. If the sequence is geometric, find the common ratio.

  a. $2,4,6,8,10,\ldots$ b. $\dfrac{2}{3}, \dfrac{4}{9}, \dfrac{8}{27}, \dfrac{16}{81}, \dfrac{32}{243}, \ldots$ c. $12, -6, 3, -\dfrac{3}{2}, \dfrac{3}{4}$

505

Section 13.3

Section 13.3 Objective 3 Finding the General Term or a Specific Term of a Geometric Sequence

Write the formula for the general term of a geometric sequence:

$a_n = $ _____

Work through Example 3 and show all work below.

Find the general term of each geometric sequence.

a. $12, -6, 3, -\dfrac{3}{2}, \dfrac{3}{4}, \dots$

b. $\dfrac{2}{3}, \dfrac{2}{9}, \dfrac{2}{27}, \dfrac{2}{81}, \dfrac{2}{243}, \dots$

Work through the interactive video with Example 4 and show all work below.

a. Find the 7th term of the geometric sequence whose first term is 2 and whose common ratio is -3.

b. Given a geometric sequence such that $a_6 = 16$ and $a_9 = 2$, find a_{13}.

Section 13.3 Objective 4 Computing the *n*th Partial Sum of a Geometric Series

Note the summation notation for an **infinite geometric series.**

What is the ***n*th partial sum** of a series? Write the partial sum in summation notation.

What is the **Formula for the *n*th Partial Sum of a Geometric Series**?

Work through the interactive video with Example 5 and show all work below.

a. Find the sum of the series $\displaystyle\sum_{i=1}^{15} 5(-2)^{i-1}$

b. Find the 7th partial sum of the geometric series $8 + 6 + \dfrac{9}{2} + \dfrac{27}{8} + \ldots$

Section 13.3

What is the Formula for the Sum of an Infinite Geometric Series?

Work through the interactive video with Example 6 and show all work below.

a. $\displaystyle\sum_{n=1}^{\infty}\frac{1}{2}\left(\frac{2}{3}\right)^{n-1}$ b) $3-\dfrac{6}{5}+\dfrac{12}{25}-\dfrac{24}{125}+\ldots$ c) $12+18+27+\dfrac{81}{2}+\dfrac{243}{4}+\ldots$

Section 13.3 Objective 6 Applications of Geometric Sequences and Series
Work through Example 7 and show all work below.

> Suppose that you have agreed to work for Donald Trump on a particular job for 21
> days. Mr. Trump gives you two choices of payment. You can be paid $100 for the
> first day and an addition $50 per day for each subsequent day. Or, you can choose to
> be paid 1 penny for the first day with your pay doubling each subsequent day. Which
> method of payment yields the most income?

Work through Example 8 and show all work below.

> A local charity received $8,500 in charitable contributions during the month of
> January. Because of a struggling economy, it is projected that contributions will
> decline each month to 95% of the previous month's contributions. What are the
> expected contributions for the month of October? What is the total expected
> contributions that this charity can expect at the end of the year?

509

Work through the interactive video with Example 9 and show all work below.
   Every repeating decimal number is a rational number and can therefore be
   represented by the quotient of two integers. Write each of the following repeating
   decimal numbers as a quotient of two integers.

   a. $.\overline{4}$ b. $.2\overline{13}$

What is the formula for the **Amount of an Ordinary Annuity after the kth Payment**?

Work through Example 10 and show all work below.
   Chie and Ben decided to save for their newborn son Jack's college education. They
   decided to invest $200 every 3 months in an investment earning 8% interest
   compounded quarterly. How much is this investment worth after 18 years?

Section 13.4 Guided Notebook

Section 13.4 The Binomial Theorem
Work through TTK # 1
Work through Objective 1
Work through Objective 2
Work through Objective 3
Work through Objective 4

Section 13.4 The Binomial Theorem

<u>**13.4 Things To Know**</u>

1. Work through TTK #1 **Multiplying Polynomials** and show all work below.

<u>Section 13.4 Objective 1 Expanding Binomials Raised to a Power Using Pascal's Triangle</u>

Work through the interactive video with Objective 1.
Determine the expansion of $(a+b)^6$ using the Binomial Theorem and Pascal's Triangle.

$n = 0$: $(a+b)^0 =$ 1

$n = 1$: $(a+b)^1 =$ $1a + 1b$

$n = 2$: $(a+b)^2 =$ $1a^2 + 2ab + 1b^2$

$n = 3$: $(a+b)^3 =$

$n = 4$: $(a+b)^4 =$

$n = 5$: $(a+b)^5 =$

Note that the sum of the exponents of each term is equal to _____.

Section 13.4

Fill in the missing blanks for Pascal's Triangle:

```
n= 0                          1
n=1                       1       1
n=2                    1      2      1
n=3                 1     _      _      1
n=4              1     4      6      4     1
n=5           1    _      _      _     _     1
```

Note below the expansion for $(a+b)^6$:

See if you can use the Binomial Theorem and Pascal's Triangle to expand each binomial found in Example 1.

a. $(x + 2)^4$

b. $(x - 3)^5$

c. $(2x - 3y)^3$

NOTE:
The terms of the expansion of the form $(a - b)^n$ will always _____ in
sign with the sign of the first term being positive. (Hint: see page 13.4-6)

Section 13.4 Objective 2 Evaluating Binomial Coefficients

What is the **Formula for a Binomial Coefficient**?

Use the formula to work through Example 2.
        Evaluate each of the following binomial coefficients.

a. $\begin{pmatrix} 5 \\ 3 \end{pmatrix}$ b. $\begin{pmatrix} 4 \\ 1 \end{pmatrix}$ c. $\begin{pmatrix} 12 \\ 8 \end{pmatrix}$

Section 13.4 Objective 3 Expanding Binomials Raised to a Power Using the Binomial
Theorem

What is the **Binomial Theorem**?

Work through the interactive video with Example 3.

     Use the Binomial Theorem to expand each binomial.

     a. $(x-1)^8$ b. $(\sqrt{x} + y^2)^5$

<u>Section 13.4 Objective 4 Finding a Particular Term or a Particular Coefficient of a Binomial Expansion</u>

The Binomial Theorem can be used to develop a formula to find a particular term within the expansion.

Formula for the $(r+1)^{\text{st}}$ Term of a Binomial Expansion

If n is a positive integer and if $r \geq 0$, then the $(r+1)^{\text{st}}$ term of

the expansion of $(a+b)^n$ is given by

$$\binom{n}{r} a^{n-r} b^r = \frac{n!}{r! \cdot (n-r)!} a^{n-r} b^r .$$

Work through the video with Example 4 and show all work below.
Find the third term of the expansion of $(2x - 3)^{10}$.

Work through the video with Example 5 and show all work below.
Find the coefficient of x^7 in the expansion of $(x + 4)^{11}$.

Section 13.5 Guided Notebook

13.5   Mathematical Induction
  Work through TTK # 1
  Work through TTK # 2
  Work through Objective 1
  Work through Objective 2

Section 13.5 Mathematical Induction

13.5 Things To Know

1.   Work through TTK #1 **Computing the *n*th Partial Sum of an Arithmetic Series** and show all work below.

2.   Work through TTK #2 **Computing the *n*th Partial Sum of a Geometric Series** and show all work below.

Section 13.5 Objective 1 Writing Mathematical Statements

Work through Example 1 and take notes below.

Given the mathematical statement $S_n : 1 + 2 + 3 + \ldots + n = \dfrac{n(n+1)}{2}$, write the

statements S_2, S_3, S_4 and S_5, and verify that each statement is true.

Work through the video with Example 2 and take notes below.

Given the statement $S_n : 1 + 2 + 3 + \ldots + n = \dfrac{n(n+1)}{2}$, write the statements S_k and S_{k+1}

for natural numbers k and k + 1.

Section 13.5 Objective 2 Using the Principle of Mathematical Induction to Prove Statements

What is The Principle of Mathematical Induction?

The 2nd condition is called the _____ step. We assume that S_k is true for some natural number k (this is called the _____ assumption), then we must show that S_{k+1} is true.

Why is Mathematical Induction similar to knocking down a sequence of dominos? See figure 5 on page 13.5–6. (Note that condition one is related to the first domino being knocked down and condition two is related to the kth domino being knocked down AND verifying that the $(k + 1)$ domino will be knocked down.)

Work through the video with Example 3 and take notes below.

Prove that the mathematical statement S_n: $1 + 2 + 3 + \ldots + n = \dfrac{n(n+1)}{2}$ is true for all natural numbers n.

Work through the video with Example 4 and take notes below.

Prove that the mathematical statement S_n: $1^3 + 2^3 + 3^3 + \ldots + n^3 = \dfrac{n^2(n+1)^2}{4}$ is true for all natural numbers n.

Section 13.5

Work through Example 5 and take notes below

Prove that the mathematical statement S_n: $2^n \geq 2n$ is true for all natural numbers n.

Section 13.6 Guided Notebook

Section 13.6 The Theory of Counting
    Work through Objective 1
    Work through Objective 2
    Work through Objective 3

Section 13.6 The Theory of Counting

<u>Section 13.6 Objective 1 Using the Fundamental Counting Principle</u>

What is the definition of **The Fundamental Counting Principle**?

Work through the video with Example 1 and take notes below.
    A woman wants to paint the exterior of her house using green for the shutters, peach
    for the siding, and off-white for the trim. She selects two shades of green, three
    shades of peach, and four shades of off-white that she likes. How many possible
    color schemes does this give her to choose from?

Section 13.6

Work through the video with Example 2 and take notes below.

A student is to create a password for his MyMathLab homework account. He is informed that the password must consist of seven characters. How many possible ways can he create the password if

a. The first three characters are letters followed by two digits and the last two characters can be letters or digits? (Repeated letters are allowed and letters are not case sensitive. Also, repeated digits are allowed.)

b. The first three characters are letters followed by two digits and the last two characters can be letters or digits? (Repeated letters are not allowed and the letters are case sensitive. Also, repeated digits are allowed.)

522

Work through the video with Example 3 and take notes below.

Linda has five t-shirts of different colors, three pairs of shorts of different patterns, and four pairs of shoes of different colors. She is planning a weekend trip to Pensacola Beach, Florida. In how many ways can she combine these items if

a. The orange t-shirt clashes with the red shoes?

b. The red t-shirt clashes with the striped shorts?

Section 13.6

Work through Example 4 and take notes below.

The instructor chooses a student to dim the lights and a student to shut the door (to use the overhead projector). In how many ways can she do this if a student may do both jobs and

a. Both are girls?

b. One is a boy and one is a girl?

c. At least one is a boy?

d. The door closer is a boy?

e. Neither are seniors?

f. Neither are education majors?

Work through the interactive video with Example 5 and take notes below.

Using the same demographic information as in Example 4, if the instructor chooses one student to dim the lights and one to shut the door, in how many ways can she do this if a student may **not** do both jobs and

a. Both are girls?

b. One is a boy and one is a girl?

c. At least one is a boy?

d. The door closer is a boy?

e. Neither are seniors?

f. Neither are education majors?

Section 13.6 Objective 2 Using Permutations

What is the definition of a **permutation**?

What is stated in the theorem, **Permutations Involving Distinct Objects with Repetition Allowed**?

Work through the video with Example 6 and take notes below.
How many four-digit arrangements are possible for a bicycle permutation lock if repetition of digits is allowed?

What is stated in the theorem, **Permutations Involving Distinct Objects with Repetition Not Allowed**?

Work through Example 7 and take notes below.

Four cards are drawn one at a time without replacement from a standard deck of 52 cards. In how many ways can they be drawn

a. When there are no restrictions as to what the cards are?

b. When all of the card are face cards?

Work through the video with Example 8 and take notes below.

There are nine competitors in a hurdle race.

a. In how many different ways can the nine competitors finish first, second, and third?

b. In how many different ways can all nine competitors finish the race?

Section 13.6

What is stated in the theorem, **Permutations Involving Some Objects That Are Not Distinct**?

Work through the video with Example 9 and take notes below.

How many distinguishable letter codes can be formed from the word SUCCESSFUL if every letter is used?

Section 13.6 Objective 3 Using Combinations

What is the definition of a **combination**?

Work through Example 10 and take notes below.

Suppose that you went to the library and you chose the following four American novels off the shelf: *Tom Sawyer*, *Little Women*, *The Graphs of Wrath*, and *The Great Gatsby*.

a. Suppose that you place two of these books back on the shelf. How many arrangements are there?

b. How many groups of two books chosen from these four books can you check out to take home?

What is stated in the theorem, **Combinations Involving Objects That Are Distinct**?

Work through the video with Example 11 and take notes below.

      The Louisiana lottery is a game in which six numbers are chosen from the numbers 1 to 44. Order does not matter and repetition of numbers is not allowed. How many possible sets of numbers are there in the Louisiana lottery?

Work through the interactive video with Example 12 and take notes below.

A poker hand consists of 5 cards that are dealt from a typical standard deck of 52 cards.

    a. How many total possible poker hands are there?

    b. How many possible hands are there that consist of exactly two aces and exactly two kings? (The fifth card is a non-ace and non-king.)

    c. How many possible poker hands are there that consist of exactly two queens and exactly three hearts?

Section 13.7 Guided Notebook

Section 13.7 An Introduction to Probability
    Work through Objective 1
    Work through Objective 2
    Work through Objective 3
    Work through Objective 4
    Work through Objective 5

Section 13.7 An Introduction to Probability

Section 13.7 Objective 1 Understanding Probability

What is the definition of **probability**?

What is the definition of **sample space**?

Work through the video with Example 1 and take notes below.
    State the number of elements in the sample space of the following experiments.
    a. Of the 20,000 students on a college campus, one is selected to be the mascot at the football game this weekend.

    b. One card is to be selected at random from a standard deck of 52 playing cards that has all of the diamonds removed.

    c. One marble is to be selected at random from a marble bag that contains two green marbles, nine red marbles, four white marbles, and five blue marbles.

    d. Three marbles are to be selected at random from a marble bag that contains two green marbles, nine red marbles, four white marbles, and five blue marbles.

531

Section 13.7

What is the formula to compute **The Probability of an Event**?

What are the three **Principles of Probability**?

Work through the video with Example 2 and take notes below.
      A paint-ball gun hopper contains 11 yellow balls, 6 green balls, and 3 red balls. A ball is fired from the gun. Find the probabilities of each of the following events written as a fraction in lowest terms:
        R: The ball is red

        G: The ball is green

        Y: The ball is yellow

What does it mean for events to be **mutually exclusive**?

Work through Example 3 and take notes below.

The instructor chooses one student to dim the lights and the next one to shut the door (to use the overhead projector). If a student may do both jobs, determine the following probabilities rounded to the nearest tenth of a percent as needed.

a. *P*(both are girls)

b. *P*(one is a girl and one is a boy)

c. *P*(At least one is a boy)

d. *P*(the door closer is a boy)

e. *P*(neither are seniors)

Section 13.7

What is the definition of **complement**?

What is **The Complement Rule**?

Work through Example 4 and take notes below.

> Using the same demographic information as in Example 3, suppose that the instructor chooses one student to dim the lights and one to shut the door. If a student may do both jobs, determine the probability that at least one is a boy using the complement rule. Write the probability rounded correct to the nearest tenth of a percent as needed.

Work through the interactive video with Example 5 and take notes below.

Using the same demographic information as in Example 3, suppose that the instructor chooses one student to dim the lights and one to shut the door. If a student may **not** do both jobs, determine the following probabilities rounded correct to the nearest tenth of a percent as needed.

a. *P*(both are girls)

b. *P*(one is a girl and one is a boy)

c. *P*(At least one is a boy)

d. *P*(the door closer is a boy)

e. *P*(neither are seniors)

Section 13.7

Section 13.7 Objective 2 Determining Probabilities Using the Additive Rule and Venn Diagrams

What is The Additive Rule of Probability?

Work through the video with Example 6 and take notes below.

Suppose $P(A) = 0.6$, $P(B) = 0.7$ and $P(A \cup B) = 0.9$. Find the following probability written as a decimal rounded to one decimal place as needed.

a. $P(A \cap B)$ b. $P((A \cap B)')$

What is a Venn Diagram?

Work through the video with Example 7 and take notes below.

In a building with 123 offices, 72 offices have computers in them. Forty-five offices have windows. Twenty-three offices have computers and windows. Construct a Venn diagram describing this information. If one office is chosen at random, determine the probability of the following events written as a fraction in lowest terms.

a. The office has a computer

b. The office has a window

c. The office has a computer and does not have a window

d. Office does not have computer and does not have a window

Section 13.7

Section 13.7 Objective 3 Using Combinations to Determine Probabilities

Work through the video with Example 8 and take notes below.

A club has 50 members, and you and your best friend are both members. Four people will be selected from this club to be in a promotional video. Determine each of the following probabilities rounded correct to the nearest tenth of a percent as needed.

a. P(you are chosen)

b. P(you and your best friend are chosen)

c. P(neither you nor your best friend are chosen)

d. P(at least one of you are chosen)

Work through the video with Example 9 and take notes below.

A box contains eight yellow, seven red, and six blue balls. Five balls are selected at random and the colors are noted. Determine each of the following probabilities written as a fraction in lowest terms.

a. P(all 5 balls are yellow)

b. P(exactly 2 balls are red)

c. P(at least one ball is blue)

Section 13.7

Section 13.7 Objective 4 Using Conditional Probability

What is the definition of **conditional probability**?

What are the two methods that can be used to determine conditional probability?

Work through the video with Example 10 and take notes below.

      The Venn diagram below shows that in a building with 123 offices, 72 offices have computers in them. Forty-five offices have windows to the outside of the building. Twenty-three offices have computers and windows. If one office is chosen at random, determine each of the following probabilities written as a fraction in lowest terms.

      a. $P(C|W)$

      b. $P(W|C)$

What is the **Conditional Probability Formula**?

Work through Example 11 and take notes below.

Suppose $P(A) = 0.6$, $P(B) = 0.7$ and $P(A \cup B) = 0.9$. Find the following probabilities correct to two decimal places as needed.

a. $P(B \mid A)$

b. $P(A \mid B)$

Section 13.7

Work through the video with Example 12 and take notes below.

A box contains eight yellow, seven red, and six blue balls. Five balls are selected at random and the colors are noted. Determine each of the following conditional probabilities written as a fraction reduced to lowest terms.

a. P(all yellow balls | all the same color)

b. P(exactly 2 red balls | no blue balls)

c. P(at least 1 blue ball | no red balls)

Section 13.7 Objective 5 Understanding Odds

What is the definition of **odds**?

Work through Example 13 and take notes below.

A private school sold 2100 raffle tickets and you purchased 20. What are the odds of winning the raffle?

Work through the video with Example 14 and take note below.

A paint-ball gun hopper contains 10 yellow balls, 6 green balls, and 3 red balls. A ball is fired from the gun. Find:

a. Odds of firing a red ball

b. Odds of firing a green ball

c. Odds against firing a yellow ball

Appendix A.1 Guided Notebook

Appendix A.1 Real Numbers

- ☐ Work through Objective 1
- ☐ Work through Objective 2
- ☐ Work through Objective 3
- ☐ Work through Objective 4

Appendix A.1 Real Numbers

Appendix A.1 Objective 1: Understanding the Real Number System

What is a **set**?

What is an **element**?

What is used to enclose the elements of a set?

What is used to name a set?

What is a **subset** and how is it written?

What are **Natural Numbers**? Write the first 10 natural numbers.

What is the purpose of the three dots?

What are **Whole Numbers**? Write the first 10 whole numbers.

What are **Integers**? Write down 5 positive and 5 negative integers.

What are **Rational Numbers**?

Are all decimals rational numbers? If not, which ones are rational?

Are natural numbers rational? Show why or why not.

Are whole numbers rational? Show why or why not.

545

Are integers rational numbers? Show why or why not.

What are **Irrational numbers**?   Write down 4 examples of irrational numbers.

What are **Real Numbers**?

List the symbols used to denote each of the sets of numbers and which set they represent.

Show the relationship between each of the sets of numbers discussed. Indicate which sets are subsets of another.

Work through Example 1 below.

Classify each number in the following set as a natural number, whole number, integer, rational number, irrational number and/or real number.

$$\left\{-5,-\frac{1}{3},0,\sqrt{3},1.\overline{4},2\pi,11\right\}$$

Appendix A.1 Objective 2: Writing Sets Using Set-Builder Notation and Interval Notation.

Answer the questions below for the following set $\left\{x\,|\,x>7\right\}$

1. Write down how this would be read.

2. Represent this set on a number line.

What is the difference between a solid circle and an open circle?

What is the purpose of the arrow?

Write down the table showing the different types of intervals, their graphs and how they would be written in set-builder notation.

Work through Example 2 and show work below.

Given the set sketched on the number line, a) identify the type of interval, b) write the set using set-builder notation, and c) write the set using interval notation.

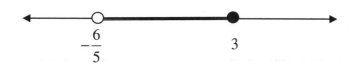

$$-\frac{6}{5} \qquad 3$$

Work Example 3, answering all the questions below, and then watch the video to check your solutions.

a. Answer the following for problem a. Write the set $[-\frac{1}{3}, \infty)$ in set-builder notation and graph the set on a number line.

    1. What does the set include?

    2. How is this written in set builder notation?

    3. Show the graph below.

    4. Why is there a solid circle at -1/3?

b. Answer the following for problem b. Write the set $\left\{ x \mid -\frac{7}{2} < x \le \pi \right\}$ in interval notation and graph the set on a number line.

    1. How is this set read?

    2. Why is there a parenthesis next to -7/2?

    3. Show the graph below.

549

Appendix A.1 Objective 3: Determining the Intersection and Union of Sets and Intervals.

What is a Union? How is it denoted?

What is an Intersection? How is it denoted?

What is the empty set and how is it denoted?

Work through Example 4 showing all work below.

$$\text{Let } A=\left\{-5,0,\frac{1}{3}11,17\right\}, B=\left\{-6,-5,4,17\right\} and \ C=\left\{-4,0,\frac{1}{4}\right\}$$

a. Find $A \cup B$ b. Find $A \cap B$ c. Find $B \cap C$

Work through Example 5, Finding the Intersection of Intervals, answering the questions below. Then check your solutions by watching the video.

a. Answer the following for problem a. $[0,\infty)\cap(-\infty,5]$

    1. Show the graph of the first interval.

    2. Show the graph of the second interval.

    3. Show the graph of the intersection. Explain the graph.

    4. Write the intersection as an interval. Why are brackets around the endpoints?

b. Answer the following for problem b. $((-\infty,-2)\cup(-2,\infty))\cap[-4,\infty)$

    1. Draw the graph of the first interval. Why is there an open dot at -2?

    2. Draw a sketch of the last interval.

3. Draw a sketch of the intersection. Why is -2 omitted? Why is there an arrow extending to the right?

4. Write the intersection in interval notation. Why is there a parenthesis next to -2?

Appendix A.1 Objective 4: Understanding Absolute Value

What is the definition of **absolute value**? How is it denoted?

Write down the 5 properties of Absolute Value.

Work through Example 6 showing all work below.

    Evaluate the following expressions involving absolute value.

a. $\left|-\sqrt{2}\right|$ b. $\left|\dfrac{-4}{16}\right|$ c. $\left|2-5\right|$

What is the **Distance between Two Real Numbers on a Number Line**? Show an example.

Work through Example 7 and show all work below.

    Find the distance between the numbers -5 and 3 using absolute value.

Appendix A.2 Guided Notebook

Appendix A.2 The Order of Operations and Algebraic Expressions

- ☐ Work through Objective 1
- ☐ Work through Objective 2
- ☐ Work through Objective 3

Appendix A.2 The Order of Operations and Algebraic Expressions

Introduction to Appendix A.2

What is an **algebraic expression**?

What are **variable**s?

Appendix A.2 Objective 1: Understanding the Properties of Real Numbers

What is the **Commutative Property of Addition**? Show an example.

What is the **Commutative Property of Multiplication**? Show an example.

What is the **Associative Property of Addition**? Show an example.

What is the **Associative Property of Multiplication**? Show an example.

What is the **Distributive Property**? Show an example.

Work through Example 1 showing all work below.

    a. Use the commutative property of addition to rewrite $11 + y$ as an equivalent expression.

    b. Use the commutative property of multiplication to rewrite $7(y + 4)$ as an equivalent expression.

    c. Use the associative property of multiplication to rewrite $8(pq)$ as an equivalent expression.

    d. Use the distributive property to multiply $11(a - 4)$.

    e. Use the distributive property to rewrite $aw + az$ as an equivalent expression.

Appendix A.2

Appendix A.2 Objective 2: Using Exponential Notation.

What is **exponential notation**?

What is b?

What is n?

Work through Example 2 showing all work below.

    a.   Rewrite 5·5·5 using exponential notation. Then identify the base and the exponent.

    b. Identify the base and the exponent of the expression $(-3)^4$, and then evaluate.

    c. Identify the base and the exponent of the expression -2^5, and then evaluate.

556

Appendix A.2 Objective 3: Using the Order of Operations to Simplify Numeric and
Algebraic Expressions.

What is the **Order of Operations**?

Work through Example 3 answering the questions below and showing all work. Then watch
the video to check your solutions.

    a.   Answer the following for problem a. $-2^3 + [3 - 5 \cdot (1 - 3)]$

        1.   What do you do to everything you're not changing?

        2. Show all steps to simplify this expression below.

Appendix A.2

Answer the following for problem b. $\dfrac{\left|2-3^3\right|+5}{5^2-4^2}$

1. Are absolute value bars a grouping symbol?

2. How do you treat the numerator and denominator when there is a large fraction bar?

3. Show all steps to simplify this expression below.

Work through Example 4 answering the questions below and showing all work. Then watch the video to check your solutions.

Evaluate the algebraic expression $-x^3 - 4x$ for $x = -2$.

1. What is the first step to evaluate this expression?

2. Show all steps to simplify this expression below.

What are **like terms**? Give an example of two terms that are like terms and two terms that are not like terms.

Work through Example 5 answering the questions below and showing all work. Then watch the video to check your solutions.

Simplify each algebraic expression.

a. Answer the following for problem a. $3x^2 - 2 + 5x^2$

1. What property allows you to rearrange the terms?

2. Show all steps to simplify this expression below using both techniques in the video.

b. Answer the following for problem b. $-3(4a - 7) + 5(3 - 5a)$

1. Show all steps to simplify this expression below.

Appendix A.3 Guided Notebook

Appendix A.3 The Laws of Exponents; Radicals

- ☐ Work through Objective 1
- ☐ Work through Objective 2
- ☐ Work through Objective 3

Appendix A.3 The Laws of Exponents; Radicals

Appendix A.3 Objective 1: Simplifying Exponential Expressions Involving Integer Exponents

Write down the **properties of positive integer exponents**.

Are they only true for positive integer exponents?

What is the **Zero Exponent Rule**?

Why can't b = 0?

What is the **Reciprocal Rule of Exponents**?

What are the 7 **Laws of Exponents**?

Appendix A.3

Work the problems in Example 1 showing all work below.

Simplify each exponential expression. Write your answers using positive exponents. Assume all variables represent positive real numbers.

a. $-3^{-4} \cdot 3^4$

b. $5z^4 \cdot (-9z^{-5})$

c. $\left(\dfrac{a^{-3}b^4c^{-6}}{2a^5b^{-4}c} \right)^{-3}$

Appendix A.3 Objective 2: Simplifying Radical Expressions.

If n is an even integer, how many nth roots does every positive real number have?

If n is an odd integer, how many nth roots does every real number have?

What symbol is used to represent the **principal nth** root of a number b?

Draw a **radical sign**.

What is n?

What is a **radical**?

562

What is the definition of the **Principal nth Root**?

Work through Example 2 showing all steps below.

Simplify each radical expression.

a. $\sqrt{144}$ b. $\sqrt[3]{\dfrac{1}{27}}$ c. $\sqrt[5]{-32a^5b^5}$ d. $\sqrt[6]{-64}$

What are the 4 **Properties of Radicals**?

563

Appendix A.3

Work through Example 3 showing all work below.

Simplify each radical expression.

a. $\sqrt{108}$

b. $\sqrt[3]{-40x^5y^3}$

c. $\sqrt[4]{16x^4y^5}$

When can radicals be added or subtracted?

Work through Example 4 showing all work below.

Simplify the radical expression. Assume all variables represent positive real numbers.

$$5\sqrt[3]{16x} + 3\sqrt[3]{2x} - \sqrt[3]{2x^4}$$

564

Watch the video with Example 5 and show all work below.

Simplify each radical expression. Assume all variables represent positive real numbers.

a. $\sqrt{192}$

b. $(3\sqrt{2}+\sqrt{5})(2\sqrt{2}-4\sqrt{5})$

c. $\sqrt[5]{2x^4y^4}\sqrt[5]{48xy^2}$

d. $8\sqrt{2x^2}-3\sqrt{28x}-2\sqrt{8x^2}$

Appendix A.3

What is rationalizing the denominator?

Watch the video with Example 6 and show all work below.

Rationalize the denominator of each expression.

a. $\sqrt{\dfrac{3}{2}}$

b. $\dfrac{\sqrt[3]{375}}{\sqrt[3]{4}}$

c. $\dfrac{7}{\sqrt{5}+\sqrt{2}}$

Appendix A.3 Objective 3: Simplifying Exponential Expressions Involving Rational Exponents

What is the definition of $b^{\frac{1}{n}}$?

566

Work through Example 7 showing all steps below.
Simplify each expression.

a. $4^{1/2}$

b. $(-27)^{1/3}$

c. $\left(\dfrac{1}{64}\right)^{\frac{1}{6}}$

What is the definition of $b^{\frac{m}{n}}$?

Write down the **Laws of Exponents**.

Appendix A.3

Work through Example 8 answering the questions below and showing all work. Then watch
the video to check your solutions.

Simplify each expression using positive exponents. Assume all variables represent
positive real numbers.

a. $(9)^{3/2}$

b. $(-32)^{-3/5}$

c. $\dfrac{\left(125x^4y^{-\frac{1}{4}}\right)^{\frac{2}{3}}}{\left(x^2y\right)^{\frac{1}{3}}}$

Appendix A.4 Guided Notebook

Appendix A.4 Polynomials

    ☐ Work through Objective 1
    ☐ Work through Objective 2
    ☐ Work through Objective 3
    ☐ Work through Objective 4

Appendix A.4 Polynomials

Appendix A.4 Objective 1: Understanding the Definition of a Polynomial

What is a **monomial**?

What is a **coefficient**?

What is the **degree**?

What is a **polynomial**?

What is a **binomial**?

What is a **trinomial**?

Appendix A.4

What is the **degree** of a polynomial?

Answer all the questions for each problem in Example 1. Show all work below.

Determine whether each algebraic expression is a polynomial. If the expression is a polynomial, state the degree.

a. $8x^3y^2 - 2$

b. $5x^5 - 3y^2 - \dfrac{6}{w}$

c. $7p^8m^4 - 5p^5m^3 + 2pm$

What is the definition of a **polynomial in one variable**?

What is **standard form** of a polynomial?

570

Appendix A.4 Objective 2: Adding and Subtracting Polynomials

What is the procedure to add or subtract polynomials?

Work the problems in Example 2 showing all steps below.

Add or subtract as indicated and express your answer in standard form.

a. $(-x^5 + 3x^4 - 9x^2 + 7x + 9) + (x^4 - 5x^5 + x^3 - 3x^2 - 8)$

b. $(-x^5 + 3x^4 - 9x^2 + 7x + 9) - (x^4 - 5x^5 + x^3 - 3x^2 - 8)$

Appendix A.4

<u>Appendix A.4 Objective 3: Multiplying Polynomials</u>

How do you find the product of two monomials in one variable?

How do you multiply a monomial by a polynomial with more than one term?

Work through Example 3 showing all work below.

Multiply: $(-2x^5)(3x^3 - 5x^2 + 1)$

Watch the video for Example 4 answering the following questions.

Find the product and express your answer in standard form.

$$(2x^2 - 3)(3x^3 - x^2 + 1)$$

What is the first step?

What is the second step?

Show all steps in detail to finish the problem.

572

The FOIL Method

What does the acronym **FOIL** stand for? Illustrate it.

Work through Example 5 showing all steps below.

Find the product: $(x + 4)(x - 4)$

Write down the 5 **special products**.

Work through the problems in Example 6. DO NOT USE FOIL! Use the Special Product formulas. Show all steps.

Find each product using a special product formula.

a. $(2x + 3)(2x - 3)$

b. $(3x + 2y)^2$

c. $(5 - 3z)^3$

Appendix A.4 Objective 4: Dividing Polynomials Using Long Division.

Write down each step of the long division process for the problem

$x+8\overline{)2x^2-7x+4}$

What is the **divisor**?

What is the **dividend**?

Show all steps of the division process below.

What is the **quotient**?

What is the **remainder**?

Appendix A.4

Watch the video with Example 7 answering the following questions.
Find the quotient and remainder when $3x^4 + x^3 + 7x + 4$ is divided by $x^2 - 1$.

1. How is the divisor rewritten?

2. How is the dividend rewritten?

3. Why do you need to put in the missing coefficients?

4. What must you do in the subtraction step?

5. Complete the problem showing all steps in detail.

Appendix A.5 Guided Notebook

Appendix A.5 Factoring Polynomials

□ Work through Objective 1
□ Work through Objective 2
□ Work through Objective 3
□ Work through Objective 4
□ Work through Objective 5

Appendix A.5 Factoring Polynomials

Factoring is the reverse of _____.

Appendix A.5 Objective 1: Factoring Out a Greatest Common Factor

Watch the video with Objective 1 and answer the following questions.

What is the distributive property?

Work the first example in the video below. (Factor: $ab + ac$)

In the second example show how he rewrites each term as a factor of primes.
(Example: $4x^2 - 6x^3 + 2x$)

577

In the example $4x^2 - 6x^3 + 2x$ what is the Greatest Common Factor?

Show the factorization below.

How can you check to see if the factorization is correct?

In the third example, what is the GCF of the constants? Of the w's? Of the y's?
(Example: $5w^2 y^3 - 15w^4 y + 20w^3 y^2$)

Continue the factorization below.

In the last example what is the GCF? (Example: $x(x^2 + 6) - 5(x^2 + 6)$)

Continue the factorization below.

Work through Example 1 showing all work below.

Factor each polynomial by factoring out the GCF.

a. $5w^2y^3 - 15w^4y + 20w^3y^2$ 　　b. $x(x^2+6) - 5(x^2+6)$

Appendix A.5 Objective 2:　Factoring by Grouping.

Watch the video with Objective 2 and answer the questions below.

When is factoring by grouping a good thing to try?

In the first 4 term example what are the two groups?

How is the first group factored?　Show the factorization below.

How is the second group factored?　Show the factorization below.

What is the common term of the two groups?

Write the complete factored form below.

In the next example, show the factorization step by step below.

Why was $-y$ the GCF of the second group?

How can you check your work?

Work the problems in Example 2 showing all steps below.

Factor each polynomial by grouping.

a. $6x^2 - 2x + 9x - 3$

b. $2x^2 + 6xw - xy - 3wy$

Appendix A.5 Objective 3: Factoring Trinomials with a Leading Coefficient Equal to One

Watch the video with Objective 3 and answer the following questions.

If c < 0 then the signs must be _____.

If c > 0 the signs must both be _____ or _____.

Write down the page of the video that is titled "Factoring Trinomials of the form $ax^2 + bx + c$."

In the first example what are the signs? Why?

What do you know about the product of the numbers in the two boxes?

What do you know about the sum of the numbers in the two boxes?

Write the factored form below.

In the next example what are the signs? Why?

What numbers belong in the boxes?

Write the factored form below.

In the last example, what are the signs? Why?

What numbers go in the boxes?

Write the factored form below.

Work through Example 3 showing all work below.

Factor each trinomial.

a. $x^2 + 8x + 15$ b. $x^2 - 2x - 24$ c. $x^2 - 12x + 32$ d. $x^2 - 2x - 6$

Appendix A.5 Objective 4: Factoring Trinomials with a Leading Coefficient Not Equal to One

Work through the animation found on page A.5-10 and answer the following questions.

What are the four steps to the **trial and error method** for factoring polynomials of the form $ax^2 + bx + c$, $a \neq 1$?

Step 1:

Step 2:

Step 3:

Step 4:

Continue to work through the animation and determine the factorization of $6x^2 + 29x + 35$ using the trial and error method.

Write down the method that will always work; **Factoring Trinomials of the form** $ax^2 + bx + c$ **, a ≠ 0, a ≠ 1.**

Work through the animation of Example 4 showing all steps below.
   Factor each trinomial.

   a. $4x^2 + 17x + 15$

   b. $12x^2 - 25x + 12$

Appendix A.5 Objective 5: Factoring Using Special Factoring Formulas
What is the **Difference of Two Squares**?

What does the note say about the sum of two squares?

What are the 2 **perfect square binomials**?

What are the formulas for a **sum or difference of 2 cubes**?

Appendix A.5

Work through the problems with Example 5. Watch the video to see if you got the correct solution.

Completely factor each expression.

a. $2x^3 + 5x^2 - 12x$

b. $5x^4 - 45x^2$

c. $8z^3x - 27y + 8z^3y - 27x$

Appendix A.6 Guided Notebook

Appendix A.6 Rational Expressions

- ☐ Work through Objective 1
- ☐ Work through Objective 2
- ☐ Work through Objective 3
- ☐ Work through Objective 4

Appendix A.6 Rational Expressions

Appendix A.6 Objective 1: Simplifying Rational Expressions

What is a **rational number**?

What is a **rational expression**?

Write the formal definition of a **Rational Expression**. Include 2 examples.

Work through Example 1 showing all work below. Watch the video to see a detailed explanation.

Simplify each rational expression.

a. $\dfrac{x^2 + x - 12}{x^2 + 9x + 20}$ b. $\dfrac{x^3 + 1}{x + 1}$ c. $\dfrac{x^2 - x - 2}{2x - x^2}$

Appendix A.6

Appendix A.6 Objective 2: Multiplying and Dividing Rational Expressions

How do you find the product of two rational expressions?

Work through Example 2 showing all steps below.

Multiply $\dfrac{2x^2+3x-2}{3x^2-2x-1}\cdot\dfrac{3x^2+4x+1}{2x^2+x-1}$

How do you find the quotient of two rational expressions?

Work through Example 3 and show work below.

Divide and simplify $\dfrac{x^3-8}{2x^2-x-6} \div \dfrac{x^2+2x+4}{6x^2+11x+3}$

Work Example 4 and show all work below. If you do not get the answer given in the book, watch the video to see the complete solution and explanation.

Perform the indicated operations and simplify.

$$\dfrac{x^2+x-6}{x^2+x-42} \cdot \dfrac{x^2+12x+35}{x^2-x-2} \div \dfrac{x+7}{x^2+8x+7}$$

Appendix A.6

Appendix A.6 Objective 3: Adding and Subtracting Rational Expressions.

To add or subtract rational numbers they must have _____.

Work through Example 5 showing all steps below.

Perform the indicated operations and simplify.

a. $\dfrac{3}{x+1} - \dfrac{2-x}{x+1}$

b. $\dfrac{3}{x^2+2x} + \dfrac{x-2}{x^2-x}$

Work through Example 6 answering the questions below. Then check your solutions by watching the video.

Answer the following questions for part a. $\dfrac{3}{x-y} - \dfrac{x+5y}{x^2-y^2}$

a. What is done first to find the LCD?

b. What form of 1 is used to modify the first fraction?

c. Continue simplifying the expression, showing all steps below.

Answer the following questions for part b. $\dfrac{x+4}{3x^2+20x+25} + \dfrac{x}{3x^2+16x+5}$

    a.   Show the steps to factor the first and second denominator.

    b. What is the LCD?

    c. What form of 1 is used to modify each fraction?

    d. Show the remaining steps to simplify the expression.

Appendix A.6

Appendix A.6 Objective 4: Simplifying Complex Rational Expressions
What is a **complex rational expression**? Give 2 examples.

What are the 3 steps of Method I for Simplifying Complex Rational Expressions?

Work through Example 7 showing all steps below.

Simplify using Method I: $\dfrac{4 - \dfrac{5}{x-1}}{\dfrac{6}{x-1} - 7}$

What are the 3 steps of Method II for Simplifying complex Rational Expressions?

Work through Example 8 and show all work below.

Simplify using Method II: $\dfrac{4 - \dfrac{5}{x-1}}{\dfrac{6}{x-1} - 7}$

Appendix A.6

Work Example 9 showing all steps below. Watch the video to see if you got the correct solution.

Simplify the complex rational expression using Method I or Method II.

$$\frac{-\dfrac{1}{x}-\dfrac{3}{x+4}}{\dfrac{2}{x^2+4x}+\dfrac{2}{x}}$$

Appendix B.1 Guided Notebook

Appendix B.1 Linear Equations

Work through Appendix B.1 TTK on page 1.1-1

Work through Objective 1

Work through Objective 2

Work through Objective 3

Work through Objective 4

Work through Objective 5

Appendix B.1 Linear Equations

Appendix B.1 Things To Know

1. Factoring Trinomials with a Leading Coefficient Equal to 1

Can you factor the polynomial $b^2 - 9b + 14$? Try working through a "You Try It" problem or refer to Appendix A.5 or watch the video.

Appendix B.1

2. Factoring Trinomials with a Leading Coefficient Not Equal to 1.

Can you factor the polynomial $15x^2 - 17x - 4$? Try working through a "You Try It" problem or refer to Appendix A.5 or watch the video.

Appendix B.1 Objective 1 Recognizing Linear Equations

What is the definition of an **algebraic expression**?

What is the definition of a **linear equation in one variable**?

596

In the Interactive Video following the definition of a linear equation in one variable, which equation is not linear? Explain why it is not linear.

Appendix B.1 Objective 2 Solving Linear Equations with Integer Coefficients

What does the term **integer coefficient** mean?

Work through Example 1 and Example 2 in your eText and take notes here:

597

Try this one on your own: Solve the following equation: $3-4(x+4)=6x-32$ and see if you can get an answer of $x=\dfrac{19}{10}$. You might want to try a "You Try It" problem now.

Appendix B.1 Objective 3 Solving Linear Equations Involving Fractions

What is the definition of a **least common denominator (LCD)**?

What is the first thing to do when solving linear equations involving fractions?

Work through the video that accompanies Example 3 and write your notes here: Solve
$$\frac{1}{3}(1-x)-\frac{x+1}{2}=-2$$

Try this one on your own: Solve the following equation: $\frac{1}{5}x-\frac{1}{3}(x-4)=\frac{1}{2}(x+1)$ and

see if you can get an answer of $x=\frac{25}{19}$. You might want to try a "You Try It" problem

now.

Appendix B.1 Objective 4 Solving Linear Equations Involving Decimals

When encountering a linear equation involving decimals, how do you eliminate the decimals?

Appendix B.1

Work through the video that accompanies Example 4 and write your notes here:

Solve $.1(y-2)+.03(y-4)=.02(10)$

Try this one on your own: Solve the following equation: $0.004(9-k)+0.04(k-9)=1$ and

see if you can get an answer of $x=\dfrac{331}{9}$. You might want to try a "You Try It" problem

now.

Appendix B.1 Objective 5 Solving Equations that Lead to Linear Equations

Work through Example 5 and take notes here: Solve $3a^2 - 1 = (a+1)(3a+2)$

Work through Example 6 and take notes here: $\dfrac{2-x}{x+2} + 3 = \dfrac{4}{x+2}$

What is an **extraneous solution**?

Appendix B.1

Work through Example 7 and take notes here: Solve $\dfrac{12}{x^2+x-2} - \dfrac{x+3}{x-1} = \dfrac{1-x}{x+2}$

(What do you have to do BEFORE you find the lowest common denominator?)

Try this one on your own: Solve the following equation: $\dfrac{10}{x^2-2x} - \dfrac{4}{x} = \dfrac{4}{x-2}$ and see if you

can get an answer of $x = \dfrac{9}{4}$. You might want to try a "You Try It" problem now.

Appendix B.2 Guided Notebook

Appendix B.2 Applications of Linear Equations

☐ Work through TTK #1
☐ Work through TTK #2
☐ Work through TTK #3
☐ Work through Objective 1
☐ Work through Objective 2
☐ Work through Objective 3
☐ Work through Objective 4
☐ Work through Objective 5

Appendix B.2 Applications of Linear Equations

Work through TTK #1 You Try It and show work below.

Work through TTK #2 You Try It and show work below.

Work through TTK #3 You Try It and show work below.

Appendix B.2

Appendix B.2 Objective 1: Converting Verbal Statements into Mathematical Statements

What are some key words and phrases and how are they translated into mathematical statements?

Work through Example 1 and show all work below.

Rewrite each statement as an algebraic expression or equation.

a. 7 more than three times a number.

b. 5 less than twice a number

c. Three times the quotient of a number and 11

d. The sum of a number and 9 is 1 less than half of the number

e. The product of a number and 4 is 1 more than 8 times the difference of 10 and the number.

What is the caution statement with Example 1 part b about?

Who is George Polya and what is he known for?

Write down Polya's Guidelines for Problem Solving.

What is the Four-Step Strategy for Problem Solving?

Appendix B.2

Work through Example 2 and show all work below.

> Roger Staubach and Terry Bradshaw were both quarterbacks in the National Football League. In 1973, Staubach threw three touchdown passes more than twice the number of touchdown passes thrown by Bradshaw. If the total number of touchdown passes between Staubach and Bradshaw was 33, how many touchdown passes did each player throw?

What expression represents the number of touchdowns Terry Bradshaw threw?

What expression represents the number of touchdowns Roger Staubach threw?

What is the equation, in words, that describes the total touchdowns thrown by both quarterbacks?

What is the algebraic equation?

What is the answer to the question in the example?

Appendix B.2 Objective 3: Solving Applications Involving Decimal Equations (Money, Mixture, Interest)

Work through TTK #3 You Try It and show work below.

Work through the video with Example 3 and take notes below.

> Billy has $16.50 in his piggy bank, consisting of nickels, dimes, and quarters. Billy notices that he has 20 fewer quarters than dimes. If the number of nickels is equal to the number of quarters and dimes combined, how many of each coin does Billy have?

What is the caution statement with Step 2 about?

What is the expression that represents the number of nickels? Why does that expression need to be multiplied by .05?

What is the expression that represents the number of quarters? Why does that expression need to be multiplied by .25?

Show all steps to solve the equation below.

Answer the question in the example.

Appendix B.2

Work through the video with Example 4 and show all work below.

How many milliliters of a 70% acid solution must be mixed with 30 mL of a 40% acid solution to obtain a mixture that is 50% acid?

What does x + 30 represent?

What does .40(30) represent?

What does .70x represent?

What does .50(30 + x) represent?

Show all steps to solve the equation below.

Answer the question in the example.

Work through the video with Example 5 and show all work below.

> Kristen inherited $20,000 from her aunt Dawn Ann, with the stipulation that she invest part of the money in an account paying 4.5% simple interest and the rest in an account paying 6% simple interest locked in for 3 years. If at the end of 1 year, the total interest earned was $982.50, how much was invested at each rate?

How is simple interest calculated?

Why is t not necessary in the formula for this example?

What are we looking for in the problem?

What did the video let x =?

What is 20000 – x according to the video?

What does the expression x(.045) represent?

What does the expression (20000 – x)(.06) represent?

Write down the correct equation as seen in the video and solve.

Appendix B.2

<u>Appendix B.2 Objective 4: Solving Applied Problems Involving Distance, Rate and Time</u>

Work through the video with Example 6 and show all work below.

> Rick left his house on his scooter at 9:00 AM to go fishing. He rode his scooter at an average speed of 10 mph. At 9:15 AM his girlfriend Deb (who did not find Rick at home) pedaled after Rick on her new 10 speed bicycle at a rate of 15 mph. If Deb caught up with Rick at precisely the time they both reached the fishing hole, how far is it from Rick's house to the fishing hole? At what time did Rick and Deb arrive at the fishing hole?

What is the relationship between distance, rate and time?

Draw the grid shown in the video and complete it along with the video.

What does t represent?

Why can't Rick's time be t + 15?

610

What is the relationship between Rick's distance and Deb's distance?

Show all steps for solving the equation.

Work through Example 7 and take notes below.

> An airplane that can maintain an average velocity of 320 mph in still air is transporting smokejumpers to a forest fire. On takeoff from the airport, it encounters a headwind and takes 34 minutes to reach the jump site. The return trip from the jump site takes 30 minutes. What is the speed of the wind? How far is it from the airport to the fire?

What affect does a headwind have on the speed of the plane?

What affect does a tailwind have on the speed of the plane?

Why are the times written as 34/60 and 30/60?

Work through the solving of the equation below.

Answer the question to Example 7.

<u>Appendix B.2 Objective 5: Solving Applied Working Together Problems</u>

Work through Example 8 answering the questions below.

> Brad and Michelle decide to paint the entire upstairs of their new house. Brad can do the job by himself in 8 hours. If it took them 3 hours to paint the upstairs together, how long would it have taken Michelle to paint it by herself?

How much of the job can Brad complete in 1 hour by himself? Why?

How much of the job will be completed in one hour if Brad and Michelle work together? Why?

Explain the equation in Method 1.

In Method 2, why is the rate to complete the job $\dfrac{1}{8}+\dfrac{1}{t}$?

Be sure you can solve the equation and get a result of $\dfrac{24}{5}$. Show work below.

Answer the question in Example 8.

Appendix B.2

Work through the video with Example 9 and take notes below.

> Jim and Earl were replacing the transmission on Earl's old convertible. Earl could replace the transmission by himself in 8 hours, whereas it would take Jim 6 hours to do the same job. They worked together for 2 hours, but then Jim had to go to his job at the grocery store. How long did it take Earl to finish replacing the transmission by himself?

What equation is used in Method 2?

Draw the chart that is in the video. Complete the information.

What does the expression (2)(1/6 + 1/8) represent? Show the steps to simplify this expression below.

How do you know Earl must complete $\frac{5}{12}$ of the job by himself?

What does t represent?

Explain the equation $x\left(\dfrac{1}{8}\right) = \dfrac{5}{12}$.

Show the steps to solve the equation below.

Answer the question for Example 9.

615

Appendix B.3 Guided Notebook

Appendix B.3 Complex Numbers

Work through Appendix B.3 TTK
Work through Objective 1
Work through Objective 2
Work through Objective 3
Work through Objective 4
Work through Objective 5

Appendix B.3 Complex Numbers

__Appendix B.3 Things To Know__

1. Simplifying Radicals

Can you simply the radical expression $\sqrt{192}$? Work through the interactive video and then try working through a "You Try It" problem or refer to Appendix A.3.

THE IMAGINARY UNIT

Take notes on the video that explains the imaginary unit here:

What is the definition of the **imaginary unit**?

Appendix B.3 Objective 1 Simplifying Powers of i

Explain the cyclic nature of powers of i:

Work through Example 1 and take notes here:
   Simplify each of the following:

a. i^{43} b. i^{100} c. i^{-21}

Try this one on your own: Write the expression i^{53} as $i, -1, -i,$ or 1. You should verify that i^{53} is equivalent to i. You might want to try a "You Try It" problem now.

Appendix B.3

COMPLEX NUMBERS

What is a complex number?

Give several examples of complex numbers.

Is every real number considered a complex number? Why or why not?

Appendix B.3 Objective 2 Adding and Subtracting Complex Numbers
Watch the video, work through Example 2 and explain how to add/subtract complex numbers.

        Perform the indicated operations:

  a. $(7 - 5i) + (-2 + i)$ b. $(7 - 5i) - (-2 + i)$

Try this one on your own: Subtract: $(9-i)-(5-7i)$.

Did you get $4+6i$? You might want to try a "You Try It" problem now.

Appendix B.3 Objective 3 Multiplying Complex Numbers

Watch the video on how to multiply complex numbers and take notes here:

Example 4: Simplify $\left(\sqrt{3}-5i\right)^{2}$. Work through the video that accompanies Example 4 and write your notes here:

619

Appendix B.3

What is the definition of a **complex conjugate**?

What will **always** happen when you multiply a complex number by its complex conjugate?

Appendix B.3 Objective 4 Finding the Quotient of Complex Numbers

Watch the video, work through Example 6 and take notes here: Write the quotient in the form $a + bi$: $\dfrac{1 - 3i}{5 - 2i}$

Try this one on your own: Divide and simplify $\dfrac{3-7i}{2+i}$ and write your answer in the form $a+bi$.

You should verify that $\dfrac{3-7i}{2+i}$ is equivalent to $-\dfrac{1}{5}-\dfrac{17}{5}i$. You might want to try a "You Try It" problem now.

Appendix B.3 Objective 5 Simplifying Radicals with Negative Radicands

Work through Example 7 and write your notes here: Simplify: $\sqrt{-108}$

True or False: $\sqrt{a}\sqrt{b}=\sqrt{ab}$ for all real numbers a and b.

Work through Example 8 and write your notes here: Simplify the following expressions:

a) $\sqrt{-8}+\sqrt{-18}$ b) $\sqrt{-8}\cdot\sqrt{-18}$ c)

$\dfrac{-6+\sqrt{(-6)^2-4(2)(5)}}{2}$ d) $\dfrac{4\pm\sqrt{-12}}{4}$

Appendix B.4 Guided Notebook

Appendix B.4 Quadratic Equations
Work through Appendix B.4 TTK
Work through Objective 1
Work through Objective 2
Work through Objective 3
Work through Objective 4
Work through Objective 5

Appendix B.4 Quadratic Equations

Appendix B.4 Things To Know

Make sure that you spend some time convincing yourself that you understand each of the following objectives. You may want to do at least one "You Try It" problem for each objective before starting this section.

1. Simplifying Radicals
2. Simplifying Radicals with Negative Radicands
3. Factoring Trinomials with a Leading Coefficient Equal to 1
4. Factoring Trinomials with a Leading Coefficient Not Equal to 1.

What is the definition of a **quadratic equation in one variable**?

Appendix B.4

<u>Appendix B.4 Objective 1 Solving Quadratic Equations by Factoring and the Zero Product Property</u>

Watch the video located under Objective 1 and take notes here: (Be sure that you know and understand the **zero product property.**)

Work through Example 1: Solve $6x^2 - 17x = -12$

Appendix B.4 Objective 2 Solving Quadratic Equations Using the Square Root Property
Watch the video located just under Objective 2 and take notes on this page:

What is the square root property and when can we use it when solving quadratic equations?

Work through Example 2 in your eText (as seen in the video) and take notes here:

a) $x^2 - 16 = 0$ b) $2x^2 + 72 = 0$ c) $(x-1)^2 = 7$

Try this one on your own: Solve the following equation using the square root property $(x+1)^2 + 16 = 0$ and see if you can get an answer of $x = -1 \pm 4i$. You might want to try a "You Try It" problem now.

Appendix B.4

Appendix B.4 Objective 3 Solving Quadratic Equations by Completing the Square

Read through Objective 3 and take notes here:

Work through Example 3 and take notes here:

a) $x^2 - 12x$

b) $x^2 + 5x$

c) $x^2 - \dfrac{3}{2}x$

Work through the video that explains the 5 steps needed to complete the square and write these 5 steps here:

1.

2.

3.

4.

5.

626

Work through Example 4. Be sure to use the 5 steps listed on your previous page of notes:

Solve $3x^2 - 18x + 19 = 0$ by completing the square.

Work through Example 5. Be sure to use the 5 steps listed on your previous page of notes:

Solve $2x^2 - 10x - 6 = 0$ by completing the square.

Appendix B.4 Objective 4 Solving Quadratic Equations Using the Quadratic Formula

You have all probably seen the quadratic formula, but where does it come from? Work through the animation that derives the quadratic formula and take notes here:

Work through Example 6 and write your notes here: Solve $3x^2 + 2x - 2 = 0$.

Work through Example 7 and write your notes here: Solve $4x^2 - x + 6 = 0$.

Appendix B.4 Objective 5 Using the Discriminant to Determine the Type of Solutions of a Quadratic Equation

Watch the video located under Objective 5 and take notes here:

Appendix B.4

Work through Example 8 and take notes here: Use the discriminant to determine the number and nature of the solutions to each of the following quadratic equations:

a) $3x^2 + 2x + 2 = 0$ b) $4x^2 + 1 = 4x$

pAppendix B.5 Guided Notebook

Appendix B.5 Applications of Quadratic Equations
Work through Appendix B.5 TTK #3

Work through Appendix B.5 TTK #4

Work through Appendix B.5 TTK #5

Work through Objective 1

Work through Objective 2

Work through Objective 3

Work through Objective 4

Work through Objective 5

Appendix B.5 Applications of Quadratic Equations

Appendix B.5 Things To Know

3. Solving Quadratic Equations by Factoring and the Zero Product Property (Appendix B.4)
   How are your factoring skills? What does the **zero product property** say? Can you solve the equation $6x^2 - 7x - 3 = 0$ by factoring and by using the zero product property?

4. Solving Quadratic Equations by Completing the Square (Appendix B.4)

Explain how to solve the equation $3x^2 - 18x + 19 = 0$ by completing the square. (Watch the video!)

5. Solving Quadratic Equations Using the Quadratic Formula (Appendix B.4)

Write down the quadratic formula and solve the equation $3x^2 - 18x + 19 = 0$ using the quadratic formula.

You should get the same answer as when you solved this equation above by completing the square.

Write down the **Four-Step Strategy for Problem Solving.**

Appendix B.5 Objective 1 Solving Applications Involving Unknown Numeric Quantities
Answer the following question for Example 1.

The product of a number and 1 more than twice the number is 36. Find the two numbers.

Explain why the other number is represented by 2x + 1.

Explain why the equation is x(2x + 1) = 36.

Show all steps to solve the equation below.

633

Appendix B.5

<u>Appendix B.5 Objective 2 Using the Projectile Motion Model</u>
What is the projectile motion model seen in this objective?

Work through Example 2 taking notes here:

     A toy rocket is launched at an initial velocity of 14.7 m/s from a 49-m tall platform.
     The height h of the object at any time t seconds after launch is given by the equation
     $h = -4.9t^2 + 14.7t + 49$ When will the rocket hit the ground?

NOTE: If you encounter a quadratic equation that does not factor, remember that you
     can solve by using the quadratic formula.

Another model used to describe projectile motion (where the height is in feet and time is in
seconds) is given by $h = -16t^2 + v_0 t + h_0$ (where v_0 is the initial velocity and h_0 is initial
height above the ground).

Appendix B.5 Objective 3 Solving Geometric Applications

Work through the interactive video that accompanies Example 3 and write your notes here:

The length of a rectangle is 6 in. less than four times the width. Find the dimensions of the rectangle if the area of the rectangle is 54 in^2.

Work through Example 4 taking notes here:

Jimmy bought a new 40 in high-definition television. If the length of Jimmy's television is 8 in. longer than the width, find the width of the television. (Remember the Pythagorean Theorem: $a^2 + b^2 = c^2$)

635

Appendix B.5

Work through the video with Example 5 showing all work below.

> Kevin flew his new Cessna O-2A airplane from Jonesburg to Mountainview, a distance of 2,560 miles. The average speed for the return trip was 64 mph faster than the average outbound speed. If the total flying time for the round trip was 18 hours, what was the plane's average speed on the outbound trip from Jonesburg to Mountainview?

Draw a picture representing the trip from Jonesburg and back.

According to the video, what does r represent?

Draw the table that is in the video and complete it.

Why does the return trip have a rate of r + 64?

Explain the expressions that represent each amount of time in the chart.

636

Explain the equation that is to be solved in the video.

Show all steps to solve the equation.

Answer the question for Example 5.

Appendix B.5 Objective 5: Solving Working Together Applications

Work through the video with Example 6 and show all work below.

> Dawn can finish the monthly sales reports in 2 hours less time than it takes Adam. Working together, they were able to finish the sales reports in 8 hours. How long does it take each person to finish the monthly sales reports alone? (Round to the nearest minute.)

According to the video, what does t represent?

Why is t – 2 Dawn's time?

Draw and complete the table shown in the video.

Explain the equation in the video.

Solve this equation below, showing all steps.

Answer the question in Example 6.

Appendix B.6 Other Types of Equations

    Work through Appendix B.6 TTK #1

    Work through Appendix B.6 TTK #2

    Work through Appendix B.6 TTK #3

    Work through Appendix B.6 TTK #4

    Work through Objective 1

    Work through Objective 2

    Work through Objective 3

Appendix B.6 Other Types of Equations

<u>Appendix B.6 Things To Know</u>

1. Factoring Trinomials with a Leading Coefficient Equal to 1 (Appendix A.5)
    It is essential that you can factor trinomials....can you? Review Appendix A.5 if you need
    a refresher and/or watch the video.

2. Factoring Trinomials with a Leading Coefficient Not Equal to 1 (Appendix A.5)

   It is essential that you can factor trinomials….can you? Review Appendix A.5 if you need a refresher and/or watch the video.

3. Factoring Polynomials by Grouping (Appendix A.5)

   When we encounter a polynomial with **4 terms** such as $2x^2 + 6xw - xy - 3wy$ it is a good idea to try to factor by grouping. Watch the video from this TTK objective to see how this polynomial is factored.

Appendix B.6

4. Solving Quadratic Equations by Factoring and the Zero Product Property (Appendix B.4)
   What does the zero product property say? Can you solve a quadratic equation by factoring? Try working through a "You Try It" problem.

Appendix B.6 Objective 1 Solving Higher-Order Polynomial Equations
Watch the video that accompanies Objective 1 and solve the following two examples that appear in this video:

   Video Example 1: Solve $10x^3 - 4x^2 = 6x$

Video Example 2: Solve $x^3 - 3x^2 + 9x - 27 = 0$

Work through Example 1 and take notes here:

Find all solutions of the equation $3x^3 - 2x = -5x^2$

Appendix B.6

Work through Example 2 and take notes here:

Find all solutions of the equation $2x^3 - x^2 + 8x - 4 = 0$

Hint: $a^2 - b^2 = (a-b)(a+b)$ "Difference of Squares"

<u>Appendix B.6 Objective 2 Solving Equations That are Quadratic In Form (Disguised Quadratics)</u>

What does it mean for an equation to be "quadratic in form"?

Work through the interactive video that accompanies Example 3 and solve each equation:

Example 3a: $2x^4 - 11x^2 + 12 = 0$

Example 3b: $\left(\dfrac{1}{x-2}\right)^2 + \dfrac{2}{x-2} - 15 = 0$

Example 3c: $x^{2/3} - 9x^{1/3} + 8 = 0$ (Hint: $\left(x^a\right)^b = x^{ab}$)

Example 3d: $3x^{-2} - 5x^{-1} - 2 = 0$

Appendix B.6

<u>Appendix B.6 Objective 3 Solving Equations Involving Radicals</u>
Work through Example 4 taking notes here:

Solve $\sqrt{x-1} - 2 = x - 9$

As indicated in the video, make sure that you ALWAYS isolate the radical prior to squaring both sides of an equation that involves a square root.

What is an **extraneous solution**?

Why is it important to check your solutions when solving equations involving radicals?

Work through the video that accompanies Example 5 taking notes here:

Solve $\sqrt{2x+3} + \sqrt{x-2} = 4$

Appendix B.7 Guided Notebook

Appendix B.7 Linear Inequalities

Work through Appendix B.7 TTK #1
Work through Appendix B.7 TTK #2
Work through Objective 1
Work through Objective 2
Work through Objective 3
Work through Objective 4

Appendix B.7 Linear Inequalities

Appendix B.7 Things To Know

1. Describing Intervals of Real Numbers (Appendix A.1)

You must get familiar with **Interval Notation, Set Builder Notation, and Using a Number Line** to describe solutions. Click on Appendix A.1 to see the following summary table which describes 5 different types of intervals.

Table 1

Type of Interval and Graph	Interval Notation	Set-Builder Notation
Open interval	(a, b)	$\{x \mid a < x < b\}$
Closed interval	$[a, b]$	$\{x \mid a \leq x \leq b\}$
Half-open intervals	$(a, b]$	$\{x \mid a < x \leq b\}$
	$[a, b)$	$\{x \mid a \leq x < b\}$
Open infinite intervals	(a, ∞)	$\{x \mid x > a\}$
	$(-\infty, b)$	$\{x \mid x < b\}$
Closed infinite intervals	$[a, \infty)$	$\{x \mid x \geq a\}$
	$(-\infty, b]$	$\{x \mid x \leq b\}$

Try Appendix A.1Example 2: Given the set sketched on the number line, a) identify the type of interval, b) write the set using set-builder notation, and c) write the set using interval notation.

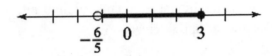

Try Appendix A.1 Example 3 and work through the video:

a) Write the set $\left[-\frac{1}{3}, \infty\right)$ in set builder notation and graph the set on a number line.

b) Write the set $\left\{ x \mid -\frac{7}{2} < x \le \pi \right\}$ in interval notation and graph the set on a number line.

Appendix B.7

Now go back to Appendix B.7 (page. B.7-1)
Appendix B.7 Things To Know

2. Understanding the Intersection and Union of Sets (Appendix A.1)
   Watch the video to see how to find the intersection and union of intervals. Take notes on the following two examples that appear in this video:

   Example a) Find the intersection: $[0,\infty) \cap (-\infty, 5]$

   Example b) Find the intersection: $\big((-\infty, -2) \cup (-2, \infty) \big) \cap [-4, \infty)$

Appendix B.7 Objective 1 Solving Linear Inequalities

What is the definition of a **linear inequality**?

650

Be sure that you are familiar with the properties of linear inequalities that are discussed in the following table.

Properties of Inequalities

Let a, b, and c be real numbers:

	Property	In Words	Example
1.	If $a < b$, then $a + c < b + c$	The same number may be added to both sides of an inequality.	$-3 < 7$ $-3 + 4 < 7 + 4$ $1 < 11$
2.	If $a < b$, then $a - c < b - c$	The same number may be subtracted from both sides of an inequality.	$9 \geq 2$ $9 - 6 \geq 2 - 6$ $3 \geq -4$
3.	For $c > 0$, if $a < b$, then $ac < bc$	Multiplying both sides of an inequality by a *positive* number *does not reverse the direction* of the inequality.	$3 > 2$ $(3)(5) > (2)(5)$ $15 > 10$
4.	For $c < 0$, if $a < b$, then $ac > bc$	Multiplying both sides of an inequality by a *negative* number *reverses the direction* of the inequality.	$3 > 2$ $(3)(-5) < (2)(-5)$ $-15 < -10$
5.	For $c > 0$, if $a < b$, then $\dfrac{a}{c} < \dfrac{b}{c}$	Dividing both sides of an inequality by a *positive* number *does not reverse the direction* of the inequality.	$6 > 4$ $\dfrac{6}{2} > \dfrac{4}{2}$ $3 > 2$
6.	For $c < 0$, if $a < b$, then $\dfrac{a}{c} > \dfrac{b}{c}$	Dividing both sides of an inequality by a *negative* number *reverses the direction* of the inequality.	$6 > 4$ $\dfrac{6}{-2} < \dfrac{4}{-2}$ $-3 < -2$

When do you reverse the direction of the inequality symbol when solving a linear inequality?

Appendix B.7

Work through Example 1 and take notes here:

Solve the inequality $-9x - 3 \geq 7 - 4x$. Graph the solution set on a number line, and express the answer in interval notation.

Work through the video that accompanies Example 2 taking notes here:

Solve the inequality $2 - 5(x - 2) < 4(3 - 2x) + 7$. Express the answer in set-builder notation.

Appendix B.7 Objective 2 Solving Three-Part Inequalities

Work through Example 3 taking notes here:

Solve the inequality $-2 \le \dfrac{2-4x}{3} < 5$. Graph the solution set on a number line, and

write the solution in set-builder notation.

Appendix B.7 Objective 3 Solving Compound Inequalities

What two words are seen in compound inequalities?

Work through Example 4 and take notes here: Solve $2x - 7 < -1$ *and* $3x + 5 \ge 3$. Graph the solution set, and write the solution in interval notation.

Appendix B.7

Work through Example 5 and take notes here:

Solve $1 - 3x \geq 7$ *or* $3x + 4 > 7$. Graph the solution set, and write the solution in interval notation.

Work through Example 6 and take notes here:

Solve $3x - 1 < -7$ *and* $4x + 1 > 9$.

654

Try this one: What is the solution to the inequality $2x < 10$ *or* $3x - 1 \geq -13$?

Did you get $(-\infty, \infty)$?

Appendix B.7 Objective 4 Solving Linear Inequality Word Problems

Work through Example 7 and take notes here:

> Suppose you rented a forklift to move a pallet with 70-lb blocks stacked on it. The forklift can carry a maximum of 2,500 lbs. If the pallet weighs 50 lbs. by itself with no blocks, how many blocks can be stacked on a pallet and lifted by the forklift?

Work through Example 8 and take notes here:

The perimeter of a rectangular fence is to be at least 80 feet and no more than 140 feet. If the width of the fence is 12 feet, what is the range of values for the length of the fence?

<u>Appendix B.8 Objective 1 Solving an Absolute Value Equation</u>
Work through the video that accompanies Example 1 and take notes here:

Solve $|1-3x|=4$.

Note that this absolute value equation is in *standard form*. Make sure that you ALWAYS get your absolute value equations into *standard form* before solving the equation.

<u>Appendix B.8 Objective 2 Solving an Absolute Value "Less Than" Inequality</u>
Work through the video that accompanies Example 2 and take notes here:

Solve $|4x-3|+2\le7$.

Appendix B.8

<u>Appendix B.8 Objective 3 Solving an Absolute Value "Greater Than" Inequality</u>
Work through the video that accompanies Example 3 and take notes here:

Solve $|5x+1| > 3$.

Carefully work through Example 4. Pay close attention to the types of solutions in this example. Note that some absolute value equations/inequalities have **no solution** and some have a solution of $(-\infty, \infty)$.

Appendix B.9 Guided Notebook

Appendix B.9 Polynomial and Rational Inequalities

    Work through Appendix B.9 TTK #1

    Work through Appendix B.9 TTK #2

    Work through Appendix B.9 TTK #3

    Work through Appendix B.9 TTK #4

    Work through Objective 1

    Work through Objective 2

Appendix B.9 Polynomial and Rational Inequalities

Appendix B.9 Things To Know

1. Factoring Trinomials with a Leading Coefficient Equal to 1 (Appendix A.5)
   It is essential that you can factor trinomials….can you? Can you factor the
trinomial $x^2 - 6x + 8$? Review Appendix A.5 if you need a refresher and/or watch the video.
You might also want to work a "You Try It" problem.

2. Factoring Trinomials with a Leading Coefficient Not Equal to 1 (Appendix A.5)

It is essential that you can factor trinomials....can you? Can you factor the trinomial $8x^2 - 2x - 15$? Review Appendix A.5 if you need a refresher and/or watch the video. You might also want to work a "You Try It" problem.

3. Factoring Polynomials by Grouping (Appendix A.5)

When we encounter a polynomial with **4 terms** such as $x^3 + x^2 - x - 1$ it is a good idea to try to factor by grouping. Can you factor this polynomial?

This polynomial factors as $(x+1)^2 (x-1)$.

4. Solving Higher-Order Polynomial Equations (Appendix B.6)

Can you solve the equation $x^3 - 3x^2 - 16x = -48$? You should get the solution set $\{3, 4, -4\}$. Try working through a "You Try It" problem.

Appendix B.9 Objective 1 Solving Polynomial Inequalities

Watch the video that accompanies Example 1 and write your notes here:

Solve the polynomial inequality $x^3 - 3x^2 + 2x \geq 0$. (Note that there is room on the next page to write down the 7 steps for solving polynomial inequalities.)

What is a **boundary point**?

What is a **test value**?

Appendix B.9

Write down the 7 Steps for Solving Polynomial Inequalities

1.

2.

3.

4.

5.

6.

7.

Follow these 7 steps to solve the polynomial inequality in Example 2:

Solve $x^2 + 5x < 3 - x^2$

Appendix B.9 Objective 2 Solving Rational Inequalities

What is the definition of a **rational inequality**?

Watch the video that accompanies Example 3 and write your notes here:

Solve $\dfrac{x-4}{x+1} \geq 0$. (Pay close attention to the boundary point obtained from the denominator. How will you plot this boundary point on a number line?)

Appendix B.9

Watch the video that accompanies Example 4 and write your notes here:

Solve $x > \dfrac{3}{x-2}$

Appendix C.1 Guided Notebook

Appendix C.1 Degree, Minute, Second Form and Degree Decimal Form
Work through Objective 1

Give an example of an angle written in **degree decimal form:**

What symbol is used for minutes?

What symbol is used for seconds?

Give an example of an angle written in **degree, minute, second form:**

Work through the interactive video that accompanies Example 1:

a. Convert $41.23°$ to DMS form. Round to the nearest second.

b. Convert $75°16'9''$ to DD form. Round to two decimal places.

Appendix C.2 Guided Notebook

Appendix C.2   Triangles

Appendix C.2 Triangles

Appendix C.2 Objective 1 Classifying Triangles

What does it mean for two angles or sides of a triangle to be **congruent**?

What is an **acute triangle**?

What is an **obtuse triangle**?

What is a **right triangle**?

Sketch and label an acute, obtuse, and right triangle, as seen in Figure 2.

What is a **scalene triangle**?

What is an **isosceles triangle**?

668

What is an **equilateral triangle**?

Sketch a scalene, isosceles, and equilateral triangle, as seen in Figure 3.

Work through Example 1 showing all work below.
 Classify the given triangle (seen in the eText on page C.2-5) as acute, obtuse, right, scalene, isosceles, or equilateral. State all that apply.

Appendix C.2 Objective 2 Using the Pythagorean Theorem

What is **The Pythagorean Theorem**? (Hint: See the text box on page C.2-6.)

669

Work through Example 2 and show all work below.

      Use the Pythagorean Theorem to find the length of the missing side of the given right triangle (as seen on page C.2-7 of the eText).

Work through the video with Example 3 and show all work below.

      A Major League baseball "diamond" is really a square. The distance between each consecutive base is 90 feet. What is the distance between home plate and second base? Round to two decimal places.

Appendix C.2 Objective 3 Understanding Similar Triangles

What is the definition of **similar triangles**?

What are the **Properties of Similar Triangles**?

1.

2.

Work through the video accompanying Example 4 showing all work below.
Triangles ABC and XYZ (as seen on page C.2-11 of the eText) are similar. Find the
lengths of the missing sides of triangle ABC.

What is the definition of the **Proportionality Constant of Similar Triangles**?

Appendix C.2

Work through the animation accompanying Example 5 showing all work below.
    The triangles below (as seen on page C.2-15 of the eText) are similar. Find the
    proportionality constant. Then find the lengths of the missing sides.

Work through the video accompanying Example 6 showing all work below.
    The right triangles below (as seen on page C.2-16 of the eText) are similar.
    Determine the lengths of the missing sides.

Appendix C.2 Objective 4 Understanding the Special Right Triangles

Sketch and label the $\dfrac{\pi}{4}, \dfrac{\pi}{4}, \dfrac{\pi}{2}$ right triangle as seen in Figure 8.

Sketch and label the $\dfrac{\pi}{6}, \dfrac{\pi}{3}, \dfrac{\pi}{2}$ right triangle as seen in Figure 12.

Work through the interactive video with Example 7 and show all work below.
    Determine the lengths of the missing sides of each right triangle (as see on page C.2-21 of the eText).

Appendix C.2 Objective 5 Using Similar Triangles to Solve Applied Problems

Work through Example 8 and show all work below.
    The shadow of a cell tower is 80 feet long. A boy 3 feet 9 inches tall is standing next to the tower. If the boy's shadow is 6 feet long, find the height of the cell tower.

Work through the video with Example 9 and show all work below.

Two people are standing on opposite sides of a small river. One person is located at point Q, a distance of 20 feet from a bridge. The other person is standing on the southeast corner of the bridge at point P. The angle between the bridge and the line of sight from P to Q is $30°$. Use this information to determine the length of the bridge and the distance between the two people. Round your answer to two decimal places as needed.